TRANSPLUX

CYBERSECURITY
In
AVIATION

Protecting the Skies from Digital Threats

Sophia Reynolds

TABLE
OF CONTENTS

PROTECTING AIRCRAFT SYSTEMS

In an era marked by rapid technological advancements and increasing digital interconnectedness, the aviation industry stands at a crossroads. While innovations in automation, artificial intelligence, and the Internet of Things (IoT) have revolutionized air travel, they have also introduced new vulnerabilities that threaten the safety, security, and efficiency of global aviation. The potential consequences of cyberattacks on aircraft, airports, and air traffic management systems are profound, ranging from operational disruptions to catastrophic incidents that could endanger lives.

"Cybersecurity in Aviation: Protecting the Skies from Digital Threats" explores the critical importance of cybersecurity in this vital sector. As aviation relies more heavily on complex digital systems, understanding and mitigating cyber threats becomes paramount for stakeholders at all levels—airlines, airport operators, manufacturers, regulatory agencies, and passengers alike. This book delves into the unique challenges posed by cyber threats, detailing how the aviation industry can develop robust security frameworks to safeguard its infrastructure and ensure passenger safety.

Throughout this book, we will examine the evolution of technology in aviation systems, the unique vulnerabilities inherent in these technologies, and the common cyber threats that the industry faces today. We will profile recent cyber incidents to highlight lessons learned and discuss the motivations behind these attacks, as well as the profiles of threat actors. Additionally, we will explore key strategies for building a culture of cybersecurity within organizations, emphasizing the need for collaboration, continuous improvement, and innovation.

The urgency for a comprehensive approach to aviation cybersecurity has never been greater. As the industry moves forward, it is essential to recognize that cyber threats are not just IT issues; they are critical safety concerns that require the commitment of every individual and organization involved in air travel. This book aims to serve as a resource for industry professionals and stakeholders seeking to understand the complexities of cybersecurity in aviation and to equip them with the knowledge necessary to create a safer, more secure environment for all.

Together, let us embark on this journey to protect the skies from digital threats, ensuring that the aviation industry remains resilient, safe, and secure for generations to come.

Chapter 1

Introduction to Cybersecurity in Aviation

Overview of Cybersecurity's Importance in the Aviation Industry

The aviation industry stands as one of the most complex and interconnected fields in modern society, involving millions of people, vast amounts of data, and extensive global operations. This interconnectedness brings a high degree of vulnerability, making the industry a prime target for cyber threats. As technological advancements continue to transform aviation - bringing in automated systems, digital navigation, in-flight connectivity, and cloud-based operations - cybersecurity has become a critical element in ensuring safe, efficient, and secure air travel.

1. A High-Value Target for Cyber Threats

The aviation industry is an attractive target for a variety of malicious actors, including cybercriminals, hacktivists, and nation-states. The motivations range from economic gains and political influence to causing disruption and testing military and defense systems. A successful cyberattack on an aircraft or airport could have devastating consequences, including potential loss of life, massive financial losses, and severe impacts on international relations. High-profile incidents can cause public mistrust, lead to regulatory penalties, and inflict reputational damage on airlines and related entities.

2. Complex and Interconnected Systems

Aviation systems are interconnected on multiple levels. These include:

- **Aircraft Systems**: The modern aircraft relies heavily on software-driven systems for functions such as flight control, navigation, and communication.

- **Ground-Based Infrastructure**: Airports and air traffic control systems use complex IT systems for managing operations, handling baggage, processing passengers, and securing the airspace.

- **Global Supply Chains**: The aviation industry involves a vast network of suppliers and contractors, from aircraft manufacturers and maintenance providers to fuel suppliers and logistics companies. This interconnected network presents numerous entry points for potential cyber threats.

Any breach in one part of this network can have far-reaching effects across the industry, highlighting the importance of a holistic and integrated approach to cybersecurity.

3. Ensuring Passenger Safety and Trust

The safety of passengers and crew is paramount in aviation. A cyberattack targeting an aircraft's avionics or communication systems could compromise flight safety, putting lives at risk. Even attacks on non-critical systems, such as in-flight entertainment or passenger data systems, can erode passenger trust. Passengers expect a secure, private, and safe experience when they travel, and even a perceived vulnerability could have long-term impacts on customer loyalty and the industry's reputation.

4. Protection of Sensitive Data and Compliance with Regulations

Aviation handles vast amounts of sensitive data, including passenger information, financial details, and operational data such as flight schedules and cargo records. The protection of this data is not only essential for the privacy and trust of passengers but is also required by stringent data protection regulations, such as the General Data Protection Regulation (GDPR) in Europe and the California Consumer Privacy Act (CCPA) in the United States. Non-compliance with these regulations can result in significant penalties and legal liabilities, making cybersecurity essential to meet regulatory requirements and maintain business integrity.

5. Safeguarding National and Global Security

Aviation plays a critical role in national and global security. The industry is integral to the defense infrastructure of many countries, with air transport and related operations often supporting military logistics and emergency response efforts. A cyberattack on aviation systems could have repercussions beyond the commercial realm, potentially impacting defense systems, critical infrastructure, and national security. Protecting aviation systems from cyber threats is therefore a priority for governments worldwide, making cybersecurity in aviation a matter of public interest and international security.

6. Mitigating Financial and Operational Risks

Cyberattacks can cause substantial financial losses for airlines, airports, and stakeholders in the aviation industry. These losses may arise from:

- **Operational Disruptions**: A successful cyberattack could halt operations, grounding flights, delaying schedules, and causing chaos at airports. Recovery from these disruptions may take days or even weeks, impacting revenue and customer satisfaction.

- **Reputational Damage**: Trust is a fundamental part of the aviation industry. A major cyber incident can lead to significant reputational damage, affecting not only the targeted organization but potentially the industry as a whole.

- **Legal and Remediation Costs**: Cyber incidents often require substantial expenditures on legal settlements, fines, and remediation efforts. In some cases, they may also lead to costly upgrades in cybersecurity systems to prevent future breaches.

By proactively investing in cybersecurity, aviation organizations can reduce the financial impact of potential incidents and ensure continuity in operations.

7. Enhancing Resilience Against Emerging Threats

The aviation industry faces a rapidly evolving threat landscape, with new types of cyber threats emerging as technology advances. Artificial intelligence, for example, has enabled both defensive and offensive capabilities in the cyber domain, including the potential for AI-powered attacks. Moreover, as quantum computing technology develops, encryption methods currently in use may become vulnerable, requiring the industry to adopt advanced encryption and cryptographic solutions. Building cybersecurity resilience in the face of these emerging threats is essential to protect both current operations and future advancements in aviation.

8. Regulatory Pressure and Industry Standards

The growing number of cyber threats has prompted international regulatory bodies, such as the International Civil Aviation Organization (ICAO), the European Union Aviation Safety Agency (EASA), and the Federal Aviation Administration (FAA), to develop cybersecurity standards and guidelines for the industry. These regulations aim to set a global standard for cybersecurity practices in aviation, ensuring that all stakeholders—airlines, airports, manufacturers, and governments—work together to mitigate cyber risks. Compliance with these standards is not only a regulatory obligation but also a strategic imperative to maintain competitive advantage and meet customer expectations.

The aviation industry has evolved from a primarily mechanical operation to one deeply integrated with digital technologies. In the early days of aviation, the focus was on mechanical reliability and preventing physical issues, such as engine failure or structural integrity. However, as aircraft became more complex and interconnected, a shift from purely mechanical threats to digital ones began. Today, cybersecurity threats represent one of the most significant challenges to aviation safety and security. This historical context highlights the journey from early mechanical concerns to the modern digital vulnerabilities that now require constant vigilance.

1. Early Mechanical Concerns and Physical Security

In the early 20th century, aviation was primarily concerned with physical and mechanical reliability:

- **Mechanical Failures**: Aircraft were vulnerable to mechanical breakdowns and physical malfunctions, such as engine failures, structural weaknesses, and control system errors. Ensuring the reliability of these systems was paramount for safe operations.

- **Human Error**: Early aviation also dealt with risks associated with pilot error, navigational inaccuracies, and limited instrumentation, which could lead to accidents. Training and standardized procedures became central to minimizing these risks.

- **Physical Security**: Threats in this era were primarily physical, such as hijacking and sabotage, which required rigorous physical security measures in airports and on aircraft.

2. Transition to Electronic and Automated Systems

Starting in the mid-20th century, advances in electronics began to transform aviation:

- **Introduction of Radar and Communication Systems**: The use of radar and radio communication introduced new levels of safety and coordination in air travel but also introduced vulnerabilities in the form of electronic interference or signal jamming.

- **Advent of Computerized Flight Systems**: By the 1970s and 1980s, aircraft increasingly relied on computerized systems for navigation, communication, and engine management. Although this shift improved safety and efficiency, it also created points of vulnerability, as computers became essential to flight operations.

- **Automated Flight Control**: The introduction of autopilot and fly-by-wire systems, which replaced manual controls with electronic systems, allowed for unprecedented control over aircraft. However, this increased dependence on automation also meant that failures in electronic systems could lead to significant operational challenges.

3. The Rise of Networked Systems and Digital Connectivity

The 1990s and early 2000s marked the dawn of the digital age in aviation, as computers became deeply integrated into every aspect of the industry:

- **Aircraft Systems Integration**: Modern aircraft became interconnected ecosystems where avionics, flight control, and communication systems shared information in real time. This development improved operational efficiency but also created multiple points of potential cyber vulnerability.

- **Ground and Air Traffic Control Connectivity**: Enhanced data exchange between ground systems and airborne aircraft led to better traffic management and smoother operations. However, the reliance on real-time data-sharing opened up new avenues for malicious actors to potentially intercept, manipulate, or jam critical communications.

- **Passenger Connectivity**: Airlines began offering in-flight Wi-Fi and entertainment systems, giving passengers access to the internet. However, this convenience also introduced new security risks, as the network infrastructure within the aircraft could potentially serve as a conduit for cyber intrusions.

4. The Emergence of Cybersecurity Threats in Aviation

With increasing connectivity and reliance on digital technology, aviation systems became susceptible to cyber threats:

- **Cyberattack on In-Flight and Ground Systems**: Airlines, airports, and aircraft systems experienced their first significant cyberattacks in the 2000s. Incidents included attempts to access aircraft systems remotely, attacks on airline IT networks, and ransomware affecting airport operations.

- **Malware and Ransomware**: Threats like malware, including ransomware, began to affect airport operations, airline bookings, and customer databases, causing disruptions and financial losses. Cybercriminals also realized the potential for targeting aviation to achieve high-impact results, prompting an era of increased vigilance.

- **Targeted Threats from Nation-States**: As cyberwarfare grew, nation-states and advanced persistent threat (APT) groups targeted aviation infrastructure to test or exploit vulnerabilities for espionage, sabotage, and intelligence-gathering purposes. This development underscored the strategic importance of cybersecurity for national security.

5. The Current Landscape: Digital Dependency and Sophisticated Threats

Today, aviation is one of the most digitally dependent industries, with cybersecurity challenges evolving as rapidly as technology itself:

- **IoT and Cloud-Based Systems**: The use of the Internet of Things (IoT) and cloud computing in aviation has improved operational efficiency, predictive maintenance, and data sharing but has also increased the attack surface. Attackers can now exploit interconnected devices and cloud vulnerabilities, posing risks to both in-flight and ground operations.

- **Automated Threat Detection**: To combat rising threats, the industry has adopted advanced cybersecurity measures, including artificial intelligence and machine learning, to detect and mitigate cyber threats in real time. However, attackers have also used AI to develop more sophisticated attacks.

- **Quantum Computing and Encryption**: As quantum computing advances, traditional encryption methods may become vulnerable. The industry is beginning to consider quantum-resistant encryption to protect sensitive information and ensure secure communication in the future.

6. The Future of Aviation Cybersecurity: Proactive and Collaborative Defense

As cyber threats continue to grow in complexity, the aviation industry must prioritize cybersecurity at all levels:

- **Adoption of Cyber Resilience Strategies**: Moving from reactive to proactive cybersecurity strategies, such as cyber resilience frameworks, is essential. This approach focuses on continuous improvement, threat anticipation, and minimizing impact in case of a breach.

- **Industry-Wide Collaboration and Information Sharing**: Recognizing that no organization can single-handedly combat cybersecurity threats, the aviation industry increasingly relies on collaboration through industry groups, regulatory bodies, and threat intelligence sharing.

- **Regulatory Support and Standardization**: Organizations like the International Civil Aviation Organization (ICAO), the Federal Aviation Administration (FAA), and the European Union Aviation Safety Agency (EASA) are establishing standards and guidelines to foster robust cybersecurity practices across the industry.

1. Cybersecurity

- **Definition**: Cybersecurity involves protecting systems, networks, and data from digital attacks, unauthorized access, damage, or disruption. In aviation, cybersecurity covers everything from aircraft avionics to airport IT systems.

- **Importance**: Effective cybersecurity in aviation ensures the safety of passengers, secures sensitive data, and maintains operational continuity.

2. Threats and Threat Actors

- **Threats**: Potential dangers to systems, data, and operations, which can include malware, ransomware, phishing, denial-of-service (DoS), and physical access attacks.

- **Threat Actors**: Entities responsible for cyber threats. These can be individuals, groups, or organizations, often classified into types:

  - **Cybercriminals**: Primarily financially motivated, targeting aviation systems for ransom, data theft, or fraud.

  - **Nation-States**: Often targeting aviation for espionage, sabotage, or geopolitical reasons.

  - **Insiders**: Employees or contractors with access to systems who may unintentionally or deliberately cause harm.

3. Vulnerability

- **Definition**: A weakness in a system, process, or application that can be exploited by a threat actor to gain unauthorized access or cause damage.

- **Types of Vulnerabilities in Aviation**:

  - **Technical Vulnerabilities**: Flaws in software, hardware, or network configurations.

  - **Human Vulnerabilities**: Insider threats, such as lack of training or unintentional errors by personnel.

  - **Physical Vulnerabilities**: Gaps in physical security measures that allow unauthorized access to sensitive areas or equipment.

4. Risk and Risk Management

- **Risk**: The potential for a cyber threat to exploit a vulnerability, leading to loss, damage, or harm. Risk is calculated by considering the likelihood of an attack and its impact.

- **Risk Management**: The process of identifying, assessing, and mitigating risks to acceptable levels. In aviation, risk management is essential for maintaining operational safety and is often part of regulatory compliance requirements.

5. Attack Surface

- **Definition**: The sum of all points where an unauthorized user can attempt to enter or extract data from an environment. In aviation, this includes everything from aircraft systems to passenger Wi-Fi networks.

- **Importance**: A larger attack surface increases the opportunities for threat actors, so minimizing it is crucial for enhancing security.

6. Encryption

- **Definition**: The process of converting data into a coded form to prevent unauthorized access. Only those with the correct decryption key can view the original data.

- **Use in Aviation**: Protects sensitive data in communication channels, like air-to-ground communications and passenger information systems, from interception or tampering.

7. Intrusion Detection Systems (IDS) and Intrusion Prevention Systems (IPS)

- **Intrusion Detection Systems (IDS)**: Monitors networks for unusual activities and potential threats, alerting operators when malicious activity is detected.

- **Intrusion Prevention Systems (IPS)**: Actively blocks detected threats in real-time, preventing unauthorized access or attacks on aviation networks.

- **Application in Aviation**: IDS and IPS help protect both aircraft and airport networks from cyber intrusions.

8. Firewall

- **Definition**: A network security system that monitors and controls incoming and outgoing network traffic based on security rules.

- **Role in Aviation**: Firewalls prevent unauthorized access to critical systems, such as aircraft avionics, air traffic management systems, and airport networks.

9. Air-Gapped Systems

- **Definition**: Systems that are physically isolated from other networks and the internet to prevent cyber threats.

- **Use in Aviation**: Air-gapping is often applied to critical flight control systems, ensuring they are isolated from potential external cyber threats.

10. Public Key Infrastructure (PKI)

- **Definition**: A framework for managing digital certificates and public-key encryption to verify identities and secure data exchanges.

- **Application in Aviation**: PKI is used in secure communications and data exchanges, ensuring that only authorized parties can access sensitive aviation systems and data.

11. Multi-Factor Authentication (MFA)

- **Definition**: A security mechanism that requires multiple forms of verification (e.g., password plus biometric scan) to grant access to systems.

- **Importance in Aviation**: MFA is crucial for restricting access to sensitive systems, such as air traffic control networks and operational control systems, helping to prevent unauthorized access.

12. Incident Response and Disaster Recovery (IR/DR)

- **Incident Response (IR)**: A structured approach to handling cybersecurity incidents, from detection and containment to eradication and recovery.

- **Disaster Recovery (DR)**: Procedures to restore systems and operations after a significant cyberattack or data breach.

- **Relevance to Aviation**: IR/DR is essential for ensuring quick and efficient recovery from cyber incidents, minimizing operational disruption, and maintaining passenger safety.

13. Zero Trust Architecture

- **Definition**: A security model based on the principle of "never trust, always verify," which restricts access to resources regardless of whether a user is inside or outside the network.

- **Importance in Aviation**: Zero Trust Architecture protects aviation systems by enforcing strict access controls, ensuring that only verified users and devices can access sensitive areas and systems.

14. Cyber Hygiene

- **Definition**: Routine practices and policies to maintain system security, reduce vulnerabilities, and minimize the risk of cyber incidents.

- **Examples in Aviation**: Regular software updates, strong password protocols, employee training, and monitoring of network activities are all part of maintaining cyber hygiene.

15. Cyber Resilience

- **Definition**: The ability of an organization or system to continue operations despite cyberattacks or incidents, minimizing disruptions and enabling swift recovery.

- **Application in Aviation**: Cyber resilience ensures continuity of critical functions, such as flight operations and air traffic control, even when facing cyber threats, safeguarding both operational and passenger safety.

Chapter 2

The Digital Transformation of Aviation

Evolution Of Technology in Aviation Systems

The evolution of technology in aviation systems has been transformative, shifting from basic mechanical controls to advanced digital and automated systems that depend heavily on secure data exchange. This journey reflects both the growth in aviation safety and efficiency, as well as the increasing complexity of managing cybersecurity risks. Here's a look at the key stages in this evolution:

1. Early Mechanical and Analog Systems (Early 20th Century - 1950s)

- **Mechanical Controls**: Early aircraft were primarily mechanical, with manual controls for flight surfaces, throttle, and other critical systems. Pilots had direct control over the aircraft through levers and cables.

- **Basic Instrumentation**: Instruments were limited to essentials like altitude, speed, and heading indicators, with analog gauges providing feedback to pilots.

- **Communication**: Initially, aviation relied on simple radio communication between aircraft and ground stations, which was essential for coordinating flights but limited in terms of security and reach.

- **Navigation Systems**: Early navigation relied on visual cues and basic instruments, including compasses and early forms of radio-based navigation, like the "radio range."

2. Introduction of Electronic Systems and Automation (1960s - 1980s)

- **Avionics Development**: The 1960s saw the emergence of "avionics" (aviation electronics), introducing systems like radar for collision avoidance, improved radio communication, and more sophisticated navigation tools, such as VOR (VHF Omnidirectional Range).

- **Autopilot and Fly-by-Wire Systems**: Autopilot systems were introduced to assist with maintaining altitude and heading, while the concept of "fly-by-wire" — replacing mechanical linkages with electronic signals — was developed to improve control and reduce weight.

- **Navigation Enhancements**: Technologies like Instrument Landing Systems (ILS) became standard, allowing for safer, more precise landings in poor visibility. Aircraft could rely on electronic signals rather than visual cues alone, enhancing safety.

- **Air Traffic Control (ATC) Improvements**: Ground systems improved with radar-based ATC, giving controllers real-time visibility into aircraft positions and making aviation safer and more efficient.

3. Digital Transformation and Networked Systems (1990s - Early 2000s)

- **Digital Cockpits and Glass Displays**: The 1990s saw the introduction of "glass cockpits," replacing analog gauges with digital screens that displayed integrated flight data. These improved situational awareness and reduced pilot workload.

- **Flight Management Systems (FMS)**: FMS allowed pilots to program routes and manage navigation automatically, making flights more efficient and reducing the risk of human error.

- **Introduction of GPS**: The Global Positioning System revolutionized navigation, enabling precise, real-time positioning for both en route and approach phases, vastly improving accuracy over previous systems.

- **Networked Ground and In-Flight Systems**: Aircraft started using data links to communicate with ground systems for flight tracking, weather updates, and logistical support, moving away from solely voice-based communication.

- **Enhanced In-Flight Entertainment and Connectivity**: Passengers began to access Wi-Fi and in-flight entertainment (IFE), which introduced new data flows and required additional network security to prevent interference with critical avionics.

4. Integrated and Automated Systems (2010s - Present)

- **NextGen and SESAR Initiatives**: U.S. and European modernization programs like NextGen (Next Generation Air Transportation System) and SESAR (Single European Sky ATM Research) introduced systems for automated air traffic management, improving efficiency and reducing delays through data-driven solutions.

- **Internet of Things (IoT) in Aviation**: Sensors and IoT devices started to monitor aircraft systems continuously, providing data for predictive maintenance. By analyzing real-time performance metrics, airlines can anticipate repairs and reduce downtime.

- **Automated and Remote Control Capabilities**: UAVs (unmanned aerial vehicles) and drones were developed for both commercial and military applications, introducing the possibility of remote-controlled and autonomous flight, which demands secure communication and control systems.

- **Big Data and Predictive Analytics**: Airlines and manufacturers now use big data analytics to optimize flight paths, fuel efficiency, and maintenance schedules. Predictive analytics help identify potential issues before they occur, enhancing safety but also creating more data that needs to be securely managed.

- **Cloud-Based Operations**: Cloud computing has enabled airlines to manage data and coordinate operations across multiple locations, facilitating real-time updates and collaboration but also adding to cybersecurity challenges.

- **Enhanced Security and Surveillance Systems**: Technologies like biometrics, advanced surveillance, and automated screening have improved physical and cyber safety, allowing for faster processing without compromising security.

5. Emerging Technologies and Future Directions

- **Artificial Intelligence (AI) and Machine Learning (ML)**: AI and ML are being used for predictive maintenance, traffic management, and even pilot assistance, allowing systems to make data-driven decisions in real-time. AI-enabled threat detection can identify cyber anomalies quickly, but these technologies also require robust cybersecurity to prevent exploitation.

- **5G and Satellite Connectivity**: High-speed 5G networks and new satellite communication technologies promise faster, more reliable connectivity for both in-flight and ground-based operations. However, these networks are also vulnerable to interception or cyberattacks if not properly secured.

- **Quantum Computing and Post-Quantum Cryptography**: Quantum computing holds the potential to revolutionize processing capabilities, but it also threatens current encryption methods. Post-quantum cryptography is becoming critical for future-proofing data protection in aviation.

- **Autonomous and Electric Aircraft**: With advancements in autonomous flight systems and electric aircraft, aviation could see fully automated air taxis and sustainable aircraft. The integration of these technologies will require enhanced cybersecurity to protect against system hacks, data breaches, and control manipulation.

- **Blockchain for Secure Data Management**: Blockchain is being explored for secure data sharing and transaction tracking within aviation logistics and supply chain management, ensuring data integrity and transparency.

1. Internet of Things (IoT) in Aviation

- **Definition and Role**: IoT connects physical devices, such as sensors and machinery, to digital networks. In aviation, IoT-enabled sensors are deployed across aircraft systems, airport facilities, and logistics to collect and exchange data in real-time.

- **Applications in Aviation**:

  o **Predictive Maintenance**: IoT sensors monitor critical components like engines, landing gear, and hydraulic systems. Real-time data allows maintenance teams to predict failures, reducing unscheduled maintenance and preventing delays.

  o **Fuel Efficiency Monitoring**: Sensors gather data on fuel consumption and engine performance, helping airlines adjust flight paths, altitudes, and speeds to optimize fuel use and reduce emissions.

  o **Environmental and Cabin Monitoring**: Temperature, air quality, and pressure sensors in the cabin improve passenger comfort, while weather sensors help pilots and ground control make informed decisions on routing.

  o **Asset Tracking**: IoT solutions help track luggage, cargo, and ground support equipment, improving operational efficiency and reducing the chance of lost items.

- **Cybersecurity Implications**: IoT devices, often connected over multiple networks, can increase the attack surface. Securing IoT endpoints, ensuring data encryption, and isolating critical systems are essential to protect against unauthorized access.

2. Artificial Intelligence (AI) in Aviation

- **Definition and Role**: AI in aviation uses machine learning algorithms and data analytics to interpret large datasets, automate tasks, and support decision-making.

- **Applications in Aviation**:

  - **Flight Path Optimization**: AI analyzes weather patterns, traffic conditions, and fuel consumption to recommend the most efficient routes, reducing fuel costs and emissions.

  - **Predictive Maintenance**: AI algorithms analyze historical maintenance and performance data to predict equipment wear and failures, further supporting IoT-driven predictive maintenance.

  - **Passenger Experience Enhancements**: AI-driven systems power chatbots, personalized in-flight entertainment, and customer service applications that respond to passenger needs in real-time.

  - **Security Screening and Facial Recognition**: AI-driven facial recognition and behavior analysis help automate and secure passenger screening, improving airport throughput and security.

  - **Traffic Management**: AI is also being tested in air traffic control to manage flight schedules, allocate airspace, and minimize delays with higher precision.

- **Cybersecurity Implications**: AI introduces data privacy concerns as it often processes sensitive data (e.g., biometric and passenger information). Additionally, AI systems must be secured against data poisoning, where attackers might try to alter training data to influence AI predictions. Monitoring and verifying AI outputs is crucial for maintaining security.

3. Automation in Aviation

- **Definition and Role**: Automation in aviation involves using technology to perform repetitive or complex tasks without direct human intervention, from autopilot systems to fully automated ground control operations.

- **Applications in Aviation**:

  - **Autopilot and Advanced Flight Management Systems**: Autopilot technology, now commonplace, allows aircraft to operate without continuous pilot input, especially useful during long-haul flights. Modern systems use automation to monitor and adjust speed, altitude, and other parameters based on real-time data.

  - **Ground Operations and Baggage Handling**: Automated systems streamline baggage handling, boarding, and loading processes, improving accuracy and efficiency.

  - **Air Traffic Control (ATC) Automation**: ATC automation tools help manage aircraft routing and sequencing, reducing the workload on human controllers and enhancing safety.

  - **Drones and Autonomous Air Taxis**: Aviation is exploring autonomous flight for applications like cargo transport, surveillance, and even passenger air taxis, which require minimal human input during flight.

- **Cybersecurity Implications**: Increased automation requires secure, reliable control systems to prevent unauthorized access and manipulation. Autonomous systems must have safeguards to override automation when needed, and all data exchanged for automated decision-making must be authenticated and protected to avoid spoofing or interference.

4. Cloud Computing in Aviation

- **Definition and Role**: Cloud computing involves using remote servers hosted on the internet to store, manage, and process data, enabling scalable, on-demand access to computing resources. In aviation, cloud platforms centralize data, streamline operations, and support large-scale data analytics.

- **Applications in Aviation**:

  o **Data Storage and Management**: Airlines and airports use cloud platforms to store vast amounts of data, including maintenance records, passenger data, and operational metrics, enabling seamless data access and sharing.

  o **Real-Time Analytics**: Cloud-based analytics provide insights from large datasets on fuel consumption, flight patterns, and maintenance needs, allowing for faster, data-driven decisions.

  o **Enhanced Collaboration**: Cloud platforms enable collaboration across departments and locations, improving coordination between airlines, airports, and regulatory authorities.

  o **Passenger Experience**: Cloud computing powers online booking, mobile check-in, and real-time flight updates, enhancing convenience for travelers.

- **Cybersecurity Implications**: Cloud environments can be targeted for data breaches, especially with sensitive passenger and operational data at risk. Securing cloud infrastructure requires strong access controls, data encryption, and compliance with industry standards like ISO/IEC 27001. Additionally, careful management of third-party access and vendor security is essential.

How Digital Systems Interconnect in Aircraft, Airports, And Air Traffic Management

The interconnectedness of digital systems in aviation is essential for seamless operations across aircraft, airports, and air traffic management (ATM). These systems create a networked ecosystem that enhances operational efficiency, safety, and coordination among all stakeholders. Here's how these digital systems interconnect and work together:

1. Aircraft Onboard Systems

- **Avionics and Flight Control Systems**: Modern aircraft have integrated avionics and flight management systems that control navigation, communication, and performance. These systems exchange data with each other and with ground control via secure, redundant networks.

- **In-Flight Communication Systems**: Systems like ACARS (Aircraft Communications Addressing and Reporting System) enable real-time data exchange between pilots, airline operations, and air traffic controllers. This allows for updates on weather, route adjustments, and aircraft status.

- **Internet of Things (IoT) Sensors**: IoT devices monitor engines, fuel levels, landing gear, and other critical components. These sensors gather data and send it to maintenance teams and data analytics platforms, which analyze performance trends and anticipate maintenance needs.

- **Passenger and Crew Networks**: In-flight entertainment systems and passenger Wi-Fi are increasingly common. These networks are typically separated from critical avionics to ensure security. Crew members also have access to systems that provide real-time updates on weather, passenger needs, and safety information.

- **Data Transmission and Connectivity**: Satellite and 5G networks support data transmission between aircraft and ground-based systems, enabling real-time communication and data updates during flight.

2. Airport Systems

- **Air Traffic Control (ATC) Integration**: Airports host control towers and ground control systems that communicate directly with ATC networks, coordinating aircraft movements on the ground and ensuring safe takeoffs and landings. ATC systems share real-time data with pilots for smooth airport operations.

- **Baggage Handling and Security Systems**: Baggage and cargo are tracked through RFID and barcode systems that feed data into centralized airport logistics systems. Automated baggage handling systems optimize sorting, reduce delays, and enhance security through interconnected scanners and monitors.

- **Passenger Management Systems**: From check-in kiosks to boarding gates, passenger management systems handle ticketing, security screening, and boarding processes. These systems are connected to airlines and government databases for security checks, and they integrate with in-flight manifests.

- **Weather Monitoring and Emergency Systems**: Airports maintain weather monitoring systems to provide up-to-the-minute data on conditions that impact flight safety and scheduling. These systems integrate with ATC to adjust takeoff and landing times and inform both pilots and ground staff.

- **Resource Management and Optimization**: Airports use digital platforms to manage gates, fueling, cleaning, and turnaround times. These systems communicate with airlines and airport operators to coordinate resources efficiently, especially during peak travel times.

3. Air Traffic Management (ATM) Systems

- **NextGen and SESAR Initiatives**: In the U.S. and Europe, modernization programs like NextGen and SESAR are implementing automated air traffic management systems that rely on data exchange between aircraft and ground stations. These systems use satellite-based navigation and tracking, moving away from radar and increasing precision.

- **Automatic Dependent Surveillance-Broadcast (ADS-B)**: ADS-B enables aircraft to transmit their location, speed, and altitude to ground stations and other aircraft. This system is critical for monitoring aircraft positions and is integrated with ATC to maintain situational awareness.

- **Flight Information Regions (FIR) and Air Traffic Flow Management (ATFM)**: ATC and ATM centers manage aircraft movements within designated airspaces. Digital ATFM systems track and control the flow of flights to prevent congestion, manage emergencies, and ensure safe distances between aircraft.

- **Data Links for Communication and Control**: Data links, such as Controller-Pilot Data Link Communications (CPDLC), provide a direct text-based communication channel between pilots and ATC. These links are crucial for long-haul flights, enabling flight path adjustments and clearances without relying solely on voice communication.

- **Real-Time Data and Analytics**: ATM systems leverage big data analytics to anticipate traffic flow, weather disruptions, and runway availability. Predictive models help ATC make real-time decisions on routing, altitude, and speed adjustments, optimizing airspace utilization.

4. Integration Across Aircraft, Airport, and ATM Systems

- **Data Sharing and Interoperability**: Airlines, airports, and ATM systems share real-time data to create a holistic view of operations. Through interconnected networks, data is constantly exchanged between aircraft (in the air), airport terminals (on the ground), and ATC/ATM (in the control centers), enabling synchronized decision-making.

- **Collaborative Decision-Making (CDM)**: CDM is a framework that allows stakeholders—such as airlines, airports, and ATC—to make joint decisions based on shared data. It helps optimize flight schedules, allocate gates, manage delays, and ensure efficient turnarounds, reducing both costs and passenger disruptions.

- **Cybersecurity and Redundancy**: To ensure data integrity and system availability, aviation networks implement cybersecurity measures such as encryption, firewall protection, and redundancy. Interconnected systems have backup protocols to maintain functionality in the event of a cyberattack or data loss.

- **Cloud Computing for Data Storage and Analysis**: Cloud platforms support data storage and processing, allowing airlines, ATC, and airports to access shared data and analytical insights remotely. Cloud computing improves collaboration and reduces the need for physical infrastructure, though it requires robust cybersecurity to protect sensitive information.

Cybersecurity Implications of Interconnected Systems

With the high level of interconnectivity in aviation, cybersecurity is critical to protect against potential breaches that could disrupt operations or endanger safety. Key considerations include:

- **Network Segmentation**: Separating critical avionics from passenger networks (e.g., in-flight entertainment) minimizes the risk of external attacks affecting essential systems.

- **Data Encryption and Authentication**: Encrypted data exchange and secure authentication protocols ensure that data between aircraft, ATC, and airports is protected from unauthorized access and manipulation.

- **Anomaly Detection**: AI-driven anomaly detection can identify unusual patterns or potential cyber threats within these interconnected systems, providing real-time alerts for response teams.

- **Compliance and Standards**: Aviation entities must comply with cybersecurity standards like the Aviation Cybersecurity Act in the EU and FAA guidelines in the U.S. These standards mandate regular security audits, incident response protocols, and data protection practices.

Chapter 3

Identifying Cyber Threats in Aviation

Common Cyber Threats

1. Malware

- **Definition**: Malware, short for malicious software, is any software intentionally designed to cause damage to a computer system, server, or network. It can take various forms, including viruses, worms, trojan horses, spyware, and adware.

- **Types of Malware in Aviation**:

    o **Viruses**: Malicious code that attaches itself to clean files and spreads throughout a computer system, often corrupting or destroying data.

    o **Worms**: Standalone malware that replicates itself to spread to other computers, often exploiting vulnerabilities in software.

- **Trojans**: Malicious software disguised as legitimate software, used to gain unauthorized access to systems.

- **Spyware**: Software that collects user data and sends it to third parties without consent, posing risks to sensitive operational data.

- **Impact on Aviation**:

  - **Disruption of Operations**: Malware can infiltrate airline reservation systems, maintenance logs, or operational databases, leading to delays and operational chaos.

  - **Data Breaches**: Malware attacks can result in unauthorized access to sensitive information, including passenger data, financial records, and security protocols.

  - **Loss of Trust**: Incidents involving malware can lead to significant reputational damage for airlines and airports, causing customers to lose trust in their ability to protect personal and financial information.

2. Ransomware

- **Definition**: Ransomware is a type of malware that encrypts the victim's data and demands a ransom payment to restore access. Ransomware attacks are often executed through phishing emails, malicious downloads, or exploiting security vulnerabilities.

- **Notable Ransomware Attacks in Aviation**:

  - **Recent Examples**: High-profile ransomware attacks against airlines or aviation service providers have targeted critical systems, leading to severe disruptions. For instance, ransomware incidents may lock down systems used for flight scheduling, crew management, or maintenance records.

- **Impact on Aviation**:
  - **Operational Downtime**: Ransomware attacks can cripple airline operations, leading to flight cancellations, delays, and financial losses due to the inability to process bookings or access necessary operational data.
  - **Financial Extortion**: Victims may face hefty ransom demands to regain access to critical systems. Paying the ransom does not guarantee data recovery and can lead to further security vulnerabilities.
  - **Regulatory Consequences**: Airlines may face legal and regulatory repercussions if they fail to adequately protect customer data or comply with cybersecurity standards.

3. Distributed Denial of Service (DDoS) Attacks

- **Definition**: A DDoS attack involves overwhelming a target system or network with a flood of traffic from multiple compromised devices, making it unavailable to legitimate users. DDoS attacks can disrupt operations and services by consuming bandwidth and processing resources.

- **Mechanism**: Attackers often use botnets—networks of infected devices—to generate a massive volume of traffic directed at a specific target, such as an airline's website or an airport's operational system.

- **Impact on Aviation**:
  - **Service Disruption**: DDoS attacks can incapacitate airline websites, online booking systems, and customer service portals, preventing customers from

making reservations or checking in, leading to frustration and lost revenue.

- ○ **Operational Delays**: If airport operational systems (e.g., flight information displays, security systems) are targeted, it can lead to delays and increased security risks, as the systems responsible for managing passenger flow and security may be overwhelmed.

- ○ **Reputation Damage**: Repeated DDoS incidents can harm an airline's reputation, leading to decreased customer confidence in their ability to provide reliable service.

Mitigation Strategies

To defend against these common cyber threats, the aviation industry can implement a range of cybersecurity measures:

1. **Robust Security Policies**: Establish comprehensive cybersecurity policies that include guidelines for handling sensitive data, system access controls, and incident response protocols.

2. **Employee Training and Awareness**: Regular training sessions for employees on recognizing phishing attempts, handling suspicious emails, and following best cybersecurity practices.

3. **Regular Software Updates and Patch Management**: Keeping all software, operating systems, and applications up-to-date to protect against known vulnerabilities that malware and ransomware can exploit.

4. **Advanced Threat Detection Systems**: Implementing intrusion detection systems (IDS) and intrusion prevention

systems (IPS) to identify and respond to unusual activity on the network.

5. **DDoS Mitigation Tools**: Utilizing DDoS protection services that can absorb and filter malicious traffic before it reaches the target systems, ensuring continued service availability.

6. **Data Backups and Recovery Plans**: Regularly backing up critical data and maintaining robust recovery plans to minimize the impact of ransomware and other data loss incidents.

7. **Collaboration with Regulatory Authorities**: Engaging with industry regulatory bodies and cybersecurity organizations to stay informed about emerging threats and best practices.

The aviation industry faces unique vulnerabilities due to the complexity of its technological systems, the critical importance of safety and security, and the increasing reliance on digital technologies. Here's an overview of some of the unique vulnerabilities in aviation technology:

1. Legacy Systems and Equipment

- **Outdated Technology**: Many aviation systems still rely on legacy technologies that may not have been designed with modern cybersecurity threats in mind. These systems can be challenging to update or replace, leaving them susceptible to attacks.

- **Compatibility Issues**: New systems must often integrate with older equipment, which may not support modern security protocols, creating potential entry points for cyber threats.

- **Limited Security Features**: Legacy systems may lack essential security features like encryption, multi-factor authentication, and real-time monitoring, making them vulnerable to exploitation.

2. Interconnectedness of Systems

- **Complex Network of Systems**: The integration of various systems—from aircraft avionics to airport management software—creates a complex network. This interconnectedness means that a vulnerability in one system can potentially compromise others.

- **Third-Party Vendors**: The use of third-party services for maintenance, data management, or software can introduce vulnerabilities, particularly if these vendors do not adhere to strict cybersecurity standards.

- **Supply Chain Risks**: The aviation supply chain includes numerous components and systems from various manufacturers. Vulnerabilities introduced at any stage can affect the overall security of aviation operations.

3. Dependence on Wireless Communications

- **Vulnerabilities in Wireless Networks**: Many aviation systems, such as ADS-B (Automatic Dependent Surveillance–Broadcast) and in-flight communication systems, rely on wireless communication, which can be susceptible to interception and jamming.

- **In-flight Wi-Fi Networks**: Passengers accessing in-flight Wi-Fi can pose security risks. Unsecured networks can allow attackers to gain access to sensitive aircraft systems if not properly segmented from critical avionics.

4. Human Factors

- **Insider Threats**: Employees with access to sensitive systems can inadvertently or maliciously compromise security. Insider threats are particularly concerning due to the potential for sabotage or data theft.

- **Human Error**: Mistakes made during operation or maintenance can lead to vulnerabilities. For example, incorrect configuration of software or hardware can leave systems open to exploitation.

- **Lack of Cybersecurity Awareness**: A lack of training and awareness among staff can lead to poor security practices, such as weak password management or falling victim to phishing attacks.

5. Data Sensitivity and Privacy Issues

- **Sensitive Information**: Aviation systems handle a vast amount of sensitive data, including passenger information, flight plans, and maintenance records. A breach could lead to severe privacy violations and regulatory repercussions.

- **Regulatory Compliance**: Aviation companies must comply with various regulations (e.g., GDPR, HIPAA) regarding data protection. Failure to protect sensitive data can lead to significant fines and legal consequences.

6. Cyber-Physical Systems Vulnerabilities

- **Integration of IT and OT**: The convergence of Information Technology (IT) and Operational Technology (OT) systems in aviation creates new vulnerabilities. Attacks on OT systems can have immediate physical impacts on safety and operations.

- **Automated Systems**: Increasing reliance on automation and AI in aviation raises concerns about the security of algorithms and the potential for malicious manipulation. Attackers could potentially exploit vulnerabilities in these automated systems to cause chaos.

7. Regulatory and Compliance Challenges

- **Evolving Threat Landscape**: The rapid pace of technological change in aviation often outstrips the development of regulations and standards, leading to potential gaps in security.

- **Global Standards Variability**: Different countries have varying regulatory requirements for cybersecurity in aviation, leading to inconsistent security postures across the industry.

8. Emerging Technologies Risks

- **Adoption of IoT**: The increased use of IoT devices for monitoring and maintenance in aviation introduces additional attack vectors, as many IoT devices may lack robust security measures.

- **AI and Machine Learning Vulnerabilities**: As airlines and airports adopt AI for predictive analytics and operational efficiencies, vulnerabilities can arise from biased algorithms, adversarial attacks, or reliance on unverified data sources.

1. British Airways Data Breach (2018)

- **Incident Overview**: British Airways (BA) experienced a data breach that compromised the personal and financial details of approximately 380,000 customers. The attack was executed by hackers who used a technique known as "skimming" to intercept data entered on the airline's website.

- **Attack Method**: Attackers injected malicious code into BA's website, capturing sensitive information such as names, email addresses, and payment card details, including CVV codes.

- **Impact**:

  - **Financial Consequences**: The breach led to significant financial losses for BA, estimated at £20 million, including potential fines from regulatory bodies under GDPR for failing to protect customer data adequately.

  - **Reputational Damage**: The incident caused substantial reputational harm to the airline, resulting in loss of customer trust.

- **Response**: British Airways took immediate action to secure its systems and informed affected customers. The airline also invested in improving cybersecurity measures to prevent future breaches.

2. United Airlines Incident (2015)

- **Incident Overview**: In May 2015, United Airlines experienced a cybersecurity incident where attackers gained access to the airline's flight operations systems, including systems responsible for scheduling and maintenance.

- **Attack Method**: It was reported that hackers exploited vulnerabilities in the airline's network, possibly using credentials obtained from previous breaches or phishing attacks.

- **Impact**:

  - **Operational Disruption**: Although the full extent of the disruption was not disclosed, there were concerns that attackers could manipulate critical flight operations data.

  - **Data Breach Concerns**: The incident raised alarms regarding the security of sensitive operational data and passenger information.

- **Response**: United Airlines emphasized its commitment to cybersecurity, enhancing its security infrastructure and protocols to protect against future incidents.

3. Sabotage of the Germanwings Flight (2015)

- **Incident Overview**: In March 2015, Germanwings Flight 9525 was deliberately crashed into the French Alps by the co-pilot, Andreas Lubitz, leading to the deaths of all 150 people on board. While this incident primarily falls under human factors, it highlights the importance of cybersecurity in mental health assessments and monitoring.

- **Vulnerabilities Exposed**:

  - **Insider Threats**: The incident raised concerns about the potential for insider threats in aviation, including the need for stringent mental health evaluations for aviation personnel.

  - **Access Control**: Questions arose regarding access controls and the need for better monitoring of flight crew behavior and health.

- **Impact**: The incident led to changes in policies regarding cockpit security and the implementation of new measures to ensure that no single pilot could control a flight alone.

4. Norwegian Air Shuttle Cyber Attack (2020)

- **Incident Overview**: In early 2020, Norwegian Air Shuttle suffered a cyber attack that targeted its internal systems, affecting operations and customer services.

- **Attack Method**: The attack involved unauthorized access to the airline's systems, although details about the specific methods used were not disclosed.

- **Impact**:

  - **Operational Challenges**: The incident led to service disruptions and complications in customer bookings and reservations.

  - **Financial Losses**: Norwegian Air Shuttle experienced financial losses attributed to the attack and subsequent recovery efforts.

- **Response**: The airline took immediate steps to secure its systems and worked with cybersecurity experts to investigate the breach and enhance its defenses.

5. Air India Data Breach (2020)

- **Incident Overview**: In May 2021, Air India disclosed a significant data breach involving the personal information of approximately 4.5 million customers due to a cyber attack on its data processor, SITA, which manages passenger service systems.

- **Attack Method**: Attackers exploited vulnerabilities in SITA's systems, which provided services to various airlines worldwide, including Air India.

- **Impact**:

  - **Data Exposure**: The breach exposed sensitive customer data, including names, contact details, and passport information, raising concerns about identity theft and fraud.

  - **Regulatory Scrutiny**: The incident prompted scrutiny from regulators regarding data protection practices and compliance with privacy laws.

- **Response**: Air India worked with SITA and cybersecurity experts to assess the damage and strengthen security measures to prevent future breaches. The airline also notified affected customers and provided guidance on monitoring for potential misuse of their data.

6. Ransomware Attack on the Colonial Pipeline (2021)

- **Incident Overview**: Although not directly related to aviation, the ransomware attack on the Colonial Pipeline in May 2021 had significant implications for the aviation industry, as it disrupted fuel supplies to airports and airlines along the East Coast of the United States.

- **Attack Method**: The attackers employed ransomware to encrypt data and demanded a ransom for its release, leading to the shutdown of the pipeline.

- **Impact**:

  - **Fuel Shortages**: Airlines experienced fuel shortages, leading to flight delays and cancellations.

  - **Operational Disruptions**: The incident highlighted the interconnected nature of critical infrastructure and its potential impact on aviation operations.

- **Response**: The attack prompted increased scrutiny of cybersecurity practices across various industries, including aviation, leading to calls for stronger protections against ransomware and improved collaboration among sectors.

Types Of Threat Actors

In the context of cybersecurity in aviation, various types of threat actors pose significant risks, each with distinct motivations and methods. Here's an overview of the main types of threat actors: nation-states, cybercriminals, hacktivists, and insiders.

1. Nation-States

- **Definition**: Nation-state actors are government-sponsored entities or groups that engage in cyber operations to further their national interests. These operations can include espionage, sabotage, and disruption of critical infrastructure.

- **Motivations**:

  o **Espionage**: Gathering intelligence on other nations' military and technological capabilities or sensitive economic data.

  o **Political Influence**: Conducting cyber operations to undermine political stability or influence public opinion in other countries.

  o **Sabotage**: Disrupting the operations of foreign airlines or airport systems to gain a strategic advantage or retaliate against perceived threats.

- **Examples**:

  o **China**: Allegations of Chinese state-sponsored hacking targeting the aviation industry to steal trade secrets and sensitive technologies.

  o **Russia**: Cyber operations aimed at destabilizing other nations' infrastructure, including potential attacks on aviation systems during geopolitical conflicts.

- **Impact on Aviation**:
  - Nation-state cyber activities can compromise sensitive aviation data, disrupt operations, and pose threats to safety and security.

2. Cybercriminals

- **Definition**: Cybercriminals are individuals or organized groups that engage in illegal activities through the internet for financial gain. They often use sophisticated techniques to exploit vulnerabilities in systems.

- **Motivations**:
  - **Financial Gain**: Theft of credit card information, personal data, and corporate secrets for sale on the dark web or direct monetary theft through ransomware.
  - **Fraud**: Engaging in identity theft, ticket fraud, or manipulating online booking systems to defraud customers or airlines.

- **Examples**:
  - **Ransomware Attacks**: Cybercriminals targeting airline systems to encrypt data and demand ransoms for restoration, such as the ransomware incident that affected various sectors, including aviation.
  - **Phishing Schemes**: Sending fraudulent emails to employees of airlines or airports to gain access to sensitive systems.

- **Impact on Aviation**:
  - Cybercriminal activities can lead to significant financial losses, operational disruptions, and

compromise customer data, affecting trust and reputation.

3. Hacktivists

- **Definition**: Hacktivists are individuals or groups that use hacking techniques to promote political agendas or social causes. Their activities often aim to draw attention to issues or protest against entities they oppose.

- **Motivations**:

    o **Political Activism**: Targeting organizations associated with government policies, environmental issues, or social justice causes.

    o **Public Awareness**: Seeking to expose perceived wrongdoing or corruption by disrupting services or leaking sensitive information.

- **Examples**:

    o **Anonymous**: This well-known hacktivist group has previously targeted various organizations, including those in the aviation sector, to protest against government policies or corporate practices.

    o **Environmental Hacktivists**: Groups focused on environmental activism may target airlines with protests against fossil fuel use or climate change policies.

- **Impact on Aviation**:

    o While often not aimed at causing direct harm, hacktivist actions can disrupt operations, lead to data breaches, and generate negative publicity for airlines.

4. Insiders

- **Definition**: Insider threats come from individuals within an organization, such as employees, contractors, or business partners, who have legitimate access to sensitive systems and data. These individuals may act maliciously or unintentionally cause harm through negligence.

- **Motivations**:

  - **Malicious Intent**: Some insiders may seek revenge against their employer, engage in corporate espionage, or sell sensitive information to competitors.

  - **Negligence**: Employees may inadvertently compromise security through careless actions, such as weak password practices or falling for phishing scams.

- **Examples**:

  - **Disgruntled Employees**: Former or current employees with grievances may intentionally leak sensitive data or disrupt operations as an act of revenge.

  - **Negligent Actions**: Employees mishandling sensitive data, such as leaving passwords exposed or connecting personal devices to secure networks, leading to unintentional data breaches.

- **Impact on Aviation**:

  - Insider threats can lead to significant data breaches, operational disruptions, and potential safety risks. Organizations must ensure robust security protocols and employee training to mitigate these risks.

Cyberattacks in the aviation sector can be driven by a variety of motivations, reflecting the complex and multifaceted nature of this industry. Understanding these motivations is critical for developing effective cybersecurity strategies. Here are some key motivations behind aviation cyberattacks:

1. Financial Gain

- **Direct Theft**: Attackers may aim to steal credit card information, personal data, or corporate secrets that can be sold on the dark web or used for identity theft.

- **Ransomware**: Cybercriminals often deploy ransomware to encrypt systems and demand payment for the decryption keys, targeting airlines and airports for substantial financial payouts due to their reliance on operational systems.

- **Fraudulent Activities**: Cybercriminals may engage in ticket fraud, using stolen credit card information to purchase tickets and then reselling them for profit.

2. Espionage

- **Corporate Espionage**: Competitors may conduct cyber operations to steal proprietary information, research, and development data, or trade secrets, especially regarding new technologies or innovations in aircraft design and manufacturing.

- **State-Sponsored Espionage**: Nation-state actors may target aviation systems to gather intelligence on other countries' military capabilities, economic activities, or technological advancements, leveraging sensitive data obtained through cyber infiltration.

3. Political and Ideological Motives

- **Activism**: Hacktivists may launch attacks on airlines or airports to protest against specific political issues, such as environmental policies, government actions, or human rights violations. Their objective is often to draw attention to their cause rather than cause direct harm.

- **Political Disruption**: Some attackers may aim to undermine the stability of a government or create chaos during significant political events by targeting the aviation sector, which is crucial for transportation and commerce.

4. Revenge and Retaliation

- **Disgruntled Employees**: Former or current employees with grievances against their employer may exploit their access to systems to disrupt operations or leak sensitive information as an act of retaliation.

- **Corporate Rivalries**: Rival companies or organizations may engage in sabotage through cyberattacks, motivated by a desire to harm competitors' operations or reputations.

5. Terrorism

- **Attacks on Infrastructure**: Terrorist groups may target aviation systems to create fear, chaos, and disruption. Such attacks can serve as a statement against governments or societies, leveraging the high visibility of aviation incidents to amplify their message.

- **Exploitation of Vulnerabilities**: Terrorist organizations may exploit gaps in cybersecurity to access critical systems and create catastrophic outcomes, potentially leading to loss of life and significant economic damage.

6. Vulnerability Exploitation

- **Demonstration of Capabilities**: Cybercriminals or hacktivists may engage in cyberattacks to showcase their skills and capabilities, often targeting high-profile organizations in the aviation sector to gain notoriety.

- **Testing Security**: Some threat actors may conduct attacks to probe the defenses of aviation organizations, seeking to identify weaknesses they can exploit in future attacks.

7. Social Engineering

- **Manipulating Individuals**: Attackers may use social engineering tactics to exploit human vulnerabilities within aviation organizations, targeting employees through phishing schemes to gain access to sensitive systems or data.

- **Credential Theft**: Cybercriminals often employ techniques to trick employees into divulging login credentials, allowing unauthorized access to critical systems and information.

Profiling recent aviation-related cyber incidents involves examining specific cases that illustrate the various threats and vulnerabilities faced by the industry. Here are detailed profiles of notable incidents from the past few years, highlighting the methods used, impacts, and responses from affected organizations:

1. SITA Data Breach (2021)

- **Overview**: SITA, a global IT provider for the aviation industry, experienced a data breach that impacted multiple airlines, including Air India and other major carriers.

- **Method**: The attackers exploited vulnerabilities in SITA's systems, which handle passenger service systems for airlines. Sensitive data, including personal information, passport details, and credit card information of approximately 4.5 million passengers, were compromised.

- **Impact**:
    - **Data Exposure**: Airlines using SITA's services faced significant risks, as personal and sensitive information of their customers was exposed.

    - **Regulatory Scrutiny**: The incident raised concerns over compliance with data protection regulations such as GDPR, leading to investigations and potential fines for the affected airlines.

- **Response**: SITA implemented enhanced security measures, including working with cybersecurity experts to investigate the breach and bolster defenses. Airlines notified affected customers and provided guidance on monitoring for identity theft.

2. Accellion Data Breach (2021)

- **Overview**: In early 2021, vulnerabilities in Accellion's File Transfer Appliance (FTA) were exploited by cybercriminals, affecting multiple sectors, including the aviation industry.

- **Method**: Attackers exploited zero-day vulnerabilities in Accellion's FTA, gaining unauthorized access to sensitive documents and data from various organizations, including airlines.

- **Impact**:

  o **Data Leaks**: Sensitive information, including employee records and passenger data, was accessed and leaked, potentially affecting numerous airlines and their customers.

  o **Reputational Damage**: Airlines involved faced reputational harm as customers lost trust in their ability to protect sensitive data.

- **Response**: Affected organizations were urged to patch vulnerabilities and improve their cybersecurity measures. Many airlines launched investigations and enhanced their data protection protocols in response to the breach.

3. Ransomware Attack on Colonial Pipeline (2021)

- **Overview**: Although not directly targeting aviation, the ransomware attack on the Colonial Pipeline had a significant impact on fuel supply to airports and airlines along the East Coast of the United States.

- **Method**: Cybercriminals used ransomware to encrypt the pipeline's operational data and demanded a ransom for the decryption key, leading to the shutdown of the pipeline.

- **Impact**:

  - ○ **Fuel Shortages**: Airlines faced fuel shortages due to the disrupted supply chain, leading to flight delays and cancellations.

  - ○ **Operational Disruptions**: The incident highlighted vulnerabilities in critical infrastructure, emphasizing the interconnectedness of sectors like aviation and energy.

- **Response**: The attack prompted federal and state governments to enhance cybersecurity measures across various sectors, including aviation, to better protect against similar incidents.

4. United Airlines Cyber Incident (2015)

- **Overview**: United Airlines faced a cyber incident that compromised flight operations systems, potentially affecting scheduling and maintenance data.

- **Method**: Attackers gained unauthorized access to United Airlines' internal network, exploiting vulnerabilities to penetrate their systems.

- **Impact**:

  - ○ **Operational Risks**: Although the extent of the disruption was not publicly detailed, the incident raised alarms about the security of critical aviation operations data.

  - ○ **Data Breach Concerns**: Concerns arose regarding the potential for data manipulation and the safety implications of compromised flight operations information.

- **Response**: United Airlines reinforced its cybersecurity measures and worked to secure its systems, emphasizing the importance of protecting sensitive operational data.

5. British Airways Data Breach (2018)

- **Overview**: British Airways experienced a major data breach that exposed the personal and financial information of around 380,000 customers.

- **Method**: Hackers used a technique known as "skimming" to inject malicious code into the airline's website, capturing data entered by customers during the booking process.

- **Impact**:
    - **Financial Consequences**: The breach led to estimated losses of £20 million for British Airways, including potential regulatory fines under GDPR.

    - **Reputational Damage**: The incident significantly damaged customer trust and raised concerns about data security practices within the airline.

- **Response**: British Airways implemented measures to enhance its website security and customer data protection, while also notifying affected customers and advising them on identity theft prevention.

6. Norwegian Air Cyber Attack (2020)

- **Overview**: Norwegian Air Shuttle faced a cyber attack that targeted its internal systems, impacting operations and customer services.

- **Method**: While specific details on the attack method were not disclosed, it was reported that unauthorized access to the airline's systems occurred.

- **Impact**:

  - ○ **Operational Disruption**: The incident led to service disruptions and complications in customer bookings and reservations.

  - ○ **Financial Losses**: Norwegian Air experienced financial losses attributed to the attack and recovery efforts.

- **Response**: The airline took immediate action to secure its systems and worked with cybersecurity experts to investigate the breach and enhance its defenses.

Chapter 5

Aircraft Systems and Cybersecurity Vulnerabilities

Overview Of Aircraft Systems and Components

The aviation industry relies on a complex array of aircraft systems and components that work together to ensure safety, efficiency, and reliability during flight. Understanding these systems is crucial for comprehending how cybersecurity risks can impact aircraft operations. Here's an overview of the main aircraft systems and components:

1. Flight Control Systems

- **Overview**: Flight control systems are critical for managing an aircraft's flight path and ensuring stability. They include both mechanical and electronic components.

- **Components**:

    o **Primary Flight Controls**: Ailerons, elevators, and rudders control roll, pitch, and yaw.

- Secondary Flight Controls: Flaps and slats assist with lift and drag during takeoff and landing.

- Fly-by-Wire Systems: Modern aircraft use electronic systems that replace traditional mechanical controls, enhancing responsiveness and safety.

2. Navigation Systems

- **Overview**: Navigation systems guide an aircraft from its departure point to its destination, ensuring accurate positioning and route adherence.

- **Components**:

  - **Global Positioning System (GPS)**: Provides real-time location data to pilots and onboard systems.

  - **Inertial Navigation Systems (INS)**: Uses accelerometers and gyroscopes to determine an aircraft's position relative to its starting point.

  - **Automatic Direction Finder (ADF)**: Helps locate radio signals from ground stations for navigation.

3. Communication Systems

- **Overview**: Communication systems enable pilots to communicate with air traffic control (ATC) and other aircraft, ensuring safety and coordination.

- **Components**:

  - **VHF Radio**: Standard communication system used for voice communication with ATC.

  - **Transponder**: Sends information about the aircraft's location and altitude to ATC radar.

o **Satellite Communication (SATCOM)**: Provides long-range communication capabilities, particularly useful over oceans or remote areas.

4. Electrical Systems

- **Overview**: Electrical systems power various components of the aircraft, from lighting to avionics.

- **Components**:

  o **Power Generation**: Engine-driven generators produce electricity during flight.

  o **Battery Systems**: Provide power for starting engines and backup systems during emergencies.

  o **Distribution Systems**: Manage the flow of electrical power to various aircraft systems.

5. Avionics Systems

- **Overview**: Avionics systems encompass a wide range of electronic systems that support flight operations.

- **Components**:

  o **Flight Management System (FMS)**: Integrates navigation, performance, and fuel management for efficient flight operations.

  o **Electronic Flight Instrument System (EFIS)**: Displays critical flight data on screens for pilot reference.

  o **Weather Radar**: Provides information on weather conditions ahead, allowing pilots to navigate around severe weather.

6. Engine Systems

- **Overview**: Aircraft engines provide the thrust necessary for flight and are critical to performance and safety.

- **Components**:

  - **Turbofan or Turboprop Engines**: The main types of engines used in commercial aviation, each with unique operational characteristics.

  - **Fuel Systems**: Manage the storage, delivery, and measurement of fuel for the engines.

  - **Engine Control Units (ECUs)**: Monitor and control engine performance, optimizing fuel efficiency and power output.

7. Environmental Control Systems

- **Overview**: These systems manage the internal environment of the aircraft, ensuring passenger comfort and safety.

- **Components**:

  - **Air Conditioning Systems**: Regulate temperature and humidity inside the cabin.

  - **Pressurization Systems**: Maintain cabin pressure at safe levels during flight at high altitudes.

  - **Oxygen Systems**: Provide supplemental oxygen in case of cabin depressurization.

8. Landing Gear Systems

- **Overview**: Landing gear systems are essential for takeoff, landing, and ground maneuvering.

- **Components**:

  - **Main and Nose Gear**: Support the aircraft on the ground and during landings.

  - **Braking Systems**: Include hydraulic brakes that provide stopping power and anti-skid systems to prevent tire wear during landing.

  - **Retractable Gear Mechanisms**: Allow landing gear to be raised or lowered as needed.

9. Safety and Emergency Systems

- **Overview**: Safety systems ensure passenger and crew safety during flight and emergencies.

- **Components**:

  - **Emergency Exits and Slides**: Allow for quick evacuation during emergencies.

  - **Fire Detection and Suppression Systems**: Monitor for signs of fire and deploy extinguishing agents as necessary.

  - **Crash Data Recorders (Black Boxes)**: Record flight data and cockpit voice to aid in accident investigations.

The vulnerabilities in avionics, communication, and navigation systems present significant challenges for the aviation industry, especially in the context of increasing cyber threats. Here's an in-depth look at these vulnerabilities, how they can be exploited, and their potential impact on aviation safety and operations:

1. Avionics Vulnerabilities

Avionics systems are the electronic systems used in aircraft for communication, navigation, and monitoring. Vulnerabilities in these systems can arise from software flaws, outdated hardware, or inadequate security measures.

- **Software Flaws**: Many avionics systems rely on complex software algorithms. Bugs or vulnerabilities in this software can be exploited by cybercriminals to disrupt aircraft operations or manipulate flight data.

- **Outdated Systems**: Legacy systems that have not been updated can lack modern security features, making them more susceptible to attacks. For example, older avionics systems may not support encryption, making it easier for hackers to intercept data.

- **Insider Threats**: Employees with access to avionics systems can inadvertently introduce vulnerabilities through negligent behavior or malicious intent. For instance, improper handling of sensitive information or access credentials can compromise system security.

- **Integration with Commercial Off-the-Shelf (COTS) Software**: The use of COTS software in avionics can introduce vulnerabilities if these products are not adequately secured or patched, potentially providing entry points for cyberattacks.

2. Communication System Vulnerabilities

Communication systems are essential for ensuring safe interactions between pilots, air traffic control (ATC), and other aircraft. Vulnerabilities in these systems can have severe implications for flight safety.

- **Unencrypted Communications**: Many communication systems, particularly older VHF radio systems, transmit data without encryption, making them susceptible to eavesdropping. Attackers can listen in on communications, potentially leading to information manipulation or exploitation.

- **Signal Jamming**: Attackers can disrupt communication channels by jamming signals, which can create confusion and disrupt normal flight operations. This could lead to loss of communication with ATC during critical phases of flight.

- **Spoofing Attacks**: Cybercriminals can use techniques to impersonate legitimate communication channels, tricking pilots or ATC into believing they are communicating with authorized entities. This could lead to miscommunication or misinformation during flight operations.

- **Vulnerable Ground Infrastructure**: Ground-based communication systems, such as those used for ATC, can also be targeted. Attacks on these systems can disrupt communication between ATC and multiple aircraft, affecting overall air traffic safety.

3. Navigation System Vulnerabilities

Navigation systems are vital for accurately determining an aircraft's position and guiding it along its intended flight path. Vulnerabilities in these systems can lead to incorrect navigation and potential accidents.

- **GPS Spoofing**: Global Positioning System (GPS) signals can be spoofed, causing an aircraft to receive false position information. This can lead to misrouting, unauthorized landings, or, in extreme cases, controlled flight into terrain.

- **Signal Interference**: Both unintentional and malicious interference can disrupt navigation signals. This can result from environmental factors, such as buildings or terrain, or deliberate attacks that jam GPS signals.

- **Dependence on Legacy Navigation Systems**: Many aircraft still rely on older navigation systems that may not incorporate modern security protocols. These systems can be vulnerable to cyber threats due to their lack of encryption and updated security features.

- **Integration with Other Systems**: Navigation systems often interface with other aircraft systems, such as autopilot and flight management systems. Vulnerabilities in these integrated systems can lead to cascading failures, where a compromise in one system affects multiple systems.

In-flight cybersecurity concerns and scenarios have become increasingly prominent as aviation technology evolves and more aircraft systems become interconnected. With advancements in avionics, communication, and passenger connectivity, the potential for cyber threats during flight raises significant concerns. Here's an overview of key cybersecurity issues and hypothetical scenarios that illustrate these concerns:

1. Passenger Wi-Fi Networks

- **Concern**: Many airlines offer in-flight Wi-Fi services, which can create vulnerabilities if not properly secured. These networks can potentially be accessed by malicious actors who could exploit them to gain unauthorized access to aircraft systems.

- **Scenario**: A passenger connects to the in-flight Wi-Fi and discovers vulnerabilities in the network. By exploiting these weaknesses, the passenger is able to launch a man-in-the-middle attack, intercepting communication between the aircraft's systems and ground control. This results in incorrect flight data being sent, causing confusion for the pilots.

2. Interconnected Systems and IoT Devices

- **Concern**: The increasing use of Internet of Things (IoT) devices in aircraft, such as sensor networks and entertainment systems, introduces additional attack vectors that can be exploited by cybercriminals.

- **Scenario**: A hacker gains access to the cabin crew's IoT devices, which are connected to the aircraft's network. They exploit a vulnerability in the entertainment system, allowing them to escalate privileges and gain access to critical avionics systems. This could lead to unauthorized changes in flight controls or navigation data.

3. Avionics Manipulation

- **Concern**: Vulnerabilities in avionics systems can be exploited during flight, potentially leading to severe safety risks.

- **Scenario**: An attacker finds a way to access the aircraft's avionics systems through a compromised maintenance laptop that connects to the aircraft during routine checks. By altering flight control settings, the attacker causes erratic behavior in the aircraft, leading to a loss of control. Pilots must use their training to regain manual control and ensure the safety of the aircraft.

4. Communication Interception

- **Concern**: The potential for communication interception between aircraft and air traffic control can lead to misinformation or manipulation of flight paths.

- **Scenario**: Cybercriminals use sophisticated radio equipment to intercept and manipulate communications between a commercial airliner and air traffic control. They issue false instructions to the pilots, leading them off course. The pilots, relying on their ATC communication, may follow erroneous directives, creating a potential collision course with another aircraft.

5. Remote Access Threats

- **Concern**: Some aircraft systems are connected to external networks for monitoring and diagnostics. This connectivity can create avenues for remote attacks.

- **Scenario**: An airline utilizes a cloud-based maintenance system that connects directly to the aircraft's avionics. A cybercriminal exploits vulnerabilities in the cloud infrastructure and gains remote access to the aircraft's systems while it is in flight. They manipulate flight data, leading to incorrect altitude readings and creating confusion for the flight crew.

6. Insider Threats

- **Concern**: Employees with access to sensitive systems can pose significant risks, either intentionally or unintentionally.

- **Scenario**: An IT technician with access to the aircraft's onboard systems intentionally introduces malware during a routine maintenance check. The malware infects the avionics systems and begins transmitting sensitive flight data to external servers controlled by the attacker. This data could include flight plans, passenger information, and operational protocols.

7. Social Engineering Attacks

- **Concern**: Attackers may attempt to manipulate airline personnel to gain access to sensitive systems or information.

- **Scenario**: An attacker poses as a maintenance engineer and convinces a cabin crew member to provide access to the aircraft's systems for a "routine check." Once inside, they deploy malicious software that compromises the aircraft's navigation systems.

Chapter 6

Airport Cybersecurity Challenges

Cyber Threats Specific to Airport Operations

Cyber threats specific to airport operations are a growing concern as airports increasingly rely on digital technologies and interconnected systems to manage their operations. These threats can disrupt airport functions, compromise sensitive data, and endanger passenger safety. Here's an overview of the key cyber threats faced by airport operations:

1. Data Breaches

- **Overview**: Airports collect and store vast amounts of sensitive information, including passenger data, employee records, and financial transactions. Cybercriminals may target these databases to steal or compromise data.

- **Threat Example**: A hacker gains unauthorized access to an airport's customer database, extracting personal information of thousands of passengers. This data is then sold on the dark web, leading to identity theft and fraud.

2. Ransomware Attacks

- **Overview**: Ransomware can encrypt critical data and systems, rendering them inaccessible until a ransom is paid. Airports can be particularly vulnerable due to their reliance on operational technology.

- **Threat Example**: An airport's operational management system is infected with ransomware, disrupting flight schedules, baggage handling, and security processes. The airport may be forced to pay a ransom to restore access, resulting in significant operational and financial impacts.

3. Denial-of-Service (DoS) Attacks

- **Overview**: DoS attacks flood a network with excessive traffic, rendering it unusable. Airports' digital infrastructure, including websites and booking systems, can be prime targets for these attacks.

- **Threat Example**: An airport experiences a DoS attack that overwhelms its online flight information system, preventing passengers from checking flight statuses or making bookings. This can lead to confusion, missed flights, and increased customer dissatisfaction.

4. Insider Threats

- **Overview**: Employees with access to critical systems can pose risks, either through malicious intent or human error. Insider threats can lead to unauthorized access or data leaks.

- **Threat Example**: An airport employee inadvertently exposes sensitive operational data by clicking on a phishing link, allowing attackers to infiltrate the airport's systems. This breach could lead to operational disruptions and loss of sensitive information.

5. Compromised Supply Chains

- **Overview**: Airports rely on numerous vendors and third-party service providers, which can introduce vulnerabilities if their systems are compromised.

- **Threat Example**: A third-party vendor's security is breached, allowing hackers to infiltrate the airport's network through shared access points. This could lead to disruptions in services such as catering, baggage handling, or maintenance.

6. Airport Operational Technology (OT) Vulnerabilities

- **Overview**: Many airport systems, including baggage handling, air traffic management, and security systems, rely on operational technology that may not have the same security measures as IT systems.

- **Threat Example**: An attacker exploits vulnerabilities in the baggage handling system, causing misrouted bags and delays in passenger connections. Such disruptions can have a cascading effect on overall airport operations.

7. Physical Security Compromise

- **Overview**: Cyber threats can extend beyond digital networks to impact physical security systems, such as access control and surveillance.

- **Threat Example**: Cybercriminals hack into an airport's access control system, granting unauthorized personnel access to restricted areas, which could lead to potential security breaches or sabotage.

8. Social Engineering Attacks

- **Overview**: Attackers may use social engineering techniques to manipulate airport staff into divulging sensitive information or granting access to restricted systems.

- **Threat Example**: An attacker impersonates a high-ranking airport official and convinces security staff to provide access to secure areas or sensitive information, potentially leading to security vulnerabilities.

9. Interconnected System Vulnerabilities

- **Overview**: Airports increasingly rely on interconnected systems that communicate with each other, creating potential pathways for cyber threats.

- **Threat Example**: A vulnerability in the airport's public Wi-Fi network is exploited, allowing attackers to gain access to the airport's internal network, which could lead to data breaches or disruption of critical services.

Vulnerabilities in ground systems such as baggage handling, check-in, and surveillance are critical concerns for airports as they increasingly rely on digital technologies and interconnected systems to operate efficiently. These vulnerabilities can lead to significant operational disruptions, loss of sensitive data, and compromises to passenger safety and security. Below is a detailed examination of these vulnerabilities across key ground systems:

1. Baggage Handling Systems

Vulnerabilities:

- **Automation and IoT Integration**: Many baggage handling systems incorporate automated processes and IoT devices, which can create security gaps if not adequately secured.

- **Legacy Systems**: Some airports may still use outdated baggage handling technology that lacks modern cybersecurity measures, making them vulnerable to attacks.

- **Physical Access**: Unauthorized personnel gaining access to baggage handling equipment can cause physical tampering, leading to operational failures or security breaches.

Potential Threats:

- **Malware Attacks**: Cybercriminals could deploy malware that disrupts the operation of baggage sorting systems, leading to lost or misrouted luggage.

- **Data Breaches**: If hackers gain access to the baggage handling system's database, they could extract sensitive passenger information, including travel itineraries and personal details.

- **Operational Disruption**: An attack that targets the software controlling the baggage handling process could cause delays and chaos, impacting flight schedules and passenger experience.

2. Check-in Systems

Vulnerabilities:

- **Online Check-in Security**: Online check-in systems may have vulnerabilities such as weak authentication methods, making them susceptible to credential theft or account takeovers.

- **Integration with Third-party Services**: Check-in systems often interface with third-party applications (e.g., payment processors, loyalty programs), which can introduce additional risks if these external systems are compromised.

- **User Error**: Passengers or staff may fall victim to phishing attacks that impersonate check-in services, leading to unauthorized access or data loss.

Potential Threats:

- **Ransomware Attacks**: If the check-in system is targeted by ransomware, it could disrupt the entire check-in process, forcing passengers to check in manually and creating long lines and delays.

- **Data Theft**: A successful attack could result in the theft of personal and payment information from passengers, leading to identity theft and financial fraud.

- **Denial of Service (DoS)**: An attacker may initiate a DoS attack to overwhelm the check-in system, preventing passengers from checking in for their flights and causing significant operational disruptions.

3. Surveillance Systems

Vulnerabilities:

- **Inadequate Network Security**: Many surveillance systems rely on IP cameras and connected sensors, which can be vulnerable to hacking if not properly secured.

- **Outdated Software**: Surveillance software may lack timely updates and patches, leaving systems open to exploitation.

- **Access Control Weaknesses**: Poor access controls can allow unauthorized users to gain control over surveillance feeds or manipulate camera functionalities.

Potential Threats:

- **Unauthorized Access**: Hackers could gain access to surveillance systems, disabling cameras or manipulating feeds to obscure criminal activities or unauthorized access points.

- **Data Manipulation**: If surveillance footage is altered or deleted, it can compromise investigations into security incidents or create confusion about security events.

- **Physical Security Breaches**: A compromised surveillance system could lead to vulnerabilities in physical security, allowing unauthorized individuals to enter restricted areas without detection.

Protecting data and personal information in airports is critical to maintaining passenger trust, ensuring operational integrity, and complying with legal and regulatory requirements. As airports increasingly rely on digital systems to manage operations and enhance passenger experience, the importance of robust cybersecurity measures and data protection strategies cannot be overstated. Here's a comprehensive overview of the challenges and strategies for safeguarding data and personal information in airport environments:

1. Challenges in Data Protection

a. Volume of Data Collected

Airports gather vast amounts of data from various sources, including ticket purchases, check-in procedures, and passenger information. This data can include personally identifiable information (PII) such as names, addresses, passport numbers, and credit card details.

b. Interconnected Systems

The integration of various systems, such as check-in, baggage handling, security screening, and loyalty programs, creates numerous points of vulnerability that can be exploited by cybercriminals.

c. Third-party Vendors

Airports often work with third-party vendors for services like payment processing, luggage handling, and cybersecurity. These partnerships can introduce additional risks if vendors do not have strong cybersecurity practices.

d. Insider Threats

Employees with access to sensitive data can pose significant risks, either intentionally or through negligence. Insider threats can lead to data breaches or unauthorized access to critical systems.

2. Strategies for Protecting Data and Personal Information

a. Data Encryption

- **Implementation**: Use strong encryption protocols for data at rest and in transit. This ensures that even if data is intercepted or accessed without authorization, it remains unreadable.

- **Benefit**: Encryption protects sensitive information from being easily exploited in the event of a data breach.

b. Access Control Measures

- **Role-based Access**: Implement strict access control measures that limit data access based on job roles and responsibilities. Only authorized personnel should have access to sensitive information.

- **Multi-factor Authentication**: Require multi-factor authentication for systems that handle personal data to add an extra layer of security against unauthorized access.

c. Regular Security Audits and Vulnerability Assessments

- **Routine Checks**: Conduct regular security audits and vulnerability assessments to identify and address potential weaknesses in data protection practices.

- **Penetration Testing**: Engage third-party experts to conduct penetration testing to evaluate the effectiveness of security measures and identify areas for improvement.

d. Employee Training and Awareness

- **Cybersecurity Training**: Provide regular training for airport staff on cybersecurity best practices, including recognizing phishing attempts and safeguarding sensitive data.

- **Awareness Programs**: Foster a culture of security awareness, encouraging employees to report suspicious activities or potential data breaches.

e. Incident Response Planning

- **Preparedness**: Develop and maintain an incident response plan that outlines procedures for responding to data breaches or security incidents.

- **Regular Drills**: Conduct regular drills to ensure that staff are familiar with response protocols and can act swiftly in the event of a data breach.

f. Data Minimization Practices

- **Limit Data Collection**: Collect only the data necessary for specific operational purposes. Reducing the volume of sensitive information stored decreases the potential impact of a data breach.

- **Anonymization**: Where possible, use anonymization techniques to protect personal data, making it difficult to identify individuals from the data collected.

g. Compliance with Regulations

- **Adherence to Standards**: Ensure compliance with relevant data protection regulations, such as the General Data Protection Regulation (GDPR) or the California Consumer Privacy Act (CCPA).

- **Regular Reviews**: Regularly review data protection policies and practices to align with evolving legal requirements and industry standards.

h. Robust Network Security

- **Firewalls and Intrusion Detection Systems**: Implement firewalls and intrusion detection systems to monitor and protect airport networks from unauthorized access and cyber threats.

- **Segmentation**: Use network segmentation to separate sensitive systems from general operational networks, limiting the potential impact of a breach.

Chapter 7

Air Traffic Management and Cybersecurity

Overview Of Air Traffic Management (ATM) Systems

Air Traffic Management (ATM) systems are critical components of modern aviation, responsible for ensuring the safe and efficient movement of aircraft through controlled airspace. These systems encompass a variety of technologies, processes, and personnel that work together to manage air traffic, provide flight information, and ensure the safety of both aircraft and passengers. Here's a comprehensive overview of ATM systems, including their functions, components, challenges, and the role of technology in their operation.

1. Functions of Air Traffic Management

ATM systems perform several essential functions, including:

- **Air Traffic Control (ATC)**: Provides guidance and instructions to pilots during takeoff, flight, and landing, ensuring safe separation between aircraft and managing their flow through controlled airspace.

- **Flight Planning**: Assists airlines and pilots in planning flight routes, optimizing fuel efficiency, and complying with regulatory requirements.

- **Surveillance and Monitoring**: Uses radar and other technologies to track the position and movement of aircraft in real time, allowing for effective management of air traffic.

- **Communication**: Facilitates clear communication between air traffic controllers and pilots through radio transmissions, ensuring timely updates and instructions.

- **Emergency Handling**: Coordinates responses to emergencies, such as technical failures, medical emergencies, or severe weather conditions, ensuring the safety of passengers and crew.

2. Components of ATM Systems

ATM systems consist of several interconnected components that work together to ensure safe and efficient air traffic management:

a. Air Traffic Control Centers

- **Area Control Centers (ACCs)**: Manage high-altitude en-route traffic over large geographical areas. Controllers at ACCs provide instructions to pilots on altitude changes, route deviations, and traffic advisories.

- **Terminal Control Areas (TCAs)**: Oversee air traffic in the vicinity of major airports. These centers manage aircraft during their descent and approach phases of flight, ensuring safe separation as they prepare to land.

- **Approach Control Facilities**: Focus on managing aircraft during the final stages of their approach to landing, providing guidance on descent paths and spacing.

b. Surveillance Systems

- **Radar Systems**: Primary and secondary radar systems are used to detect and track aircraft. Primary radar provides information on the aircraft's position, while secondary radar (Transponder) enhances this data with additional information such as altitude and identification.

- **Automatic Dependent Surveillance-Broadcast (ADS-B)**: A technology that allows aircraft to determine their position via satellite navigation and periodically broadcast this information to ground stations and other aircraft.

c. Communication Systems

- **Voice Communication**: Radio systems facilitate communication between air traffic controllers and pilots, allowing for the exchange of critical information.

- **Data Link Communication**: Systems such as Controller-Pilot Data Link Communications (CPDLC) provide a text-based communication channel for exchanging messages related to flight information, reducing reliance on voice communication.

d. Flight Data Processing Systems

- **Automated Systems**: These systems process flight plans, manage traffic flow, and provide decision support tools for air traffic controllers. They help optimize airspace usage and reduce delays.

- **Conflict Detection and Resolution Tools**: Advanced algorithms analyze aircraft trajectories and provide alerts for potential conflicts, enabling controllers to take preventive actions.

3. Challenges in Air Traffic Management

Despite the sophistication of ATM systems, several challenges persist:

- **Increasing Air Traffic Volume**: The growing number of flights poses significant challenges for air traffic controllers, requiring more efficient traffic management and enhanced capacity.

- **Technological Integration**: Integrating new technologies and systems into existing ATM frameworks can be complex and costly, requiring careful planning and coordination.

- **Cybersecurity Threats**: As ATM systems become more digitized and interconnected, they face increasing risks from cyber threats, necessitating robust cybersecurity measures to protect sensitive data and operational integrity.

- **Adverse Weather Conditions**: Weather events can impact flight operations and complicate air traffic management, requiring real-time updates and contingency planning.

4. The Role of Technology in ATM Systems

Technology plays a crucial role in enhancing the capabilities and efficiency of ATM systems:

- **Automation**: Increasing automation in flight data processing and conflict detection helps reduce the workload on air traffic controllers and improve safety.

- **Artificial Intelligence (AI) and Machine Learning**: AI and machine learning algorithms are being explored to analyze traffic patterns, optimize flight routes, and enhance decision-making processes.

- **Advanced Surveillance Technologies**: Innovations such as satellite-based navigation and surveillance improve situational awareness and enable more precise tracking of aircraft.

- **Collaborative Decision Making (CDM)**: Enhanced data sharing among stakeholders—airlines, airports, and air traffic control—facilitates collaborative decision-making, optimizing resource utilization and improving overall efficiency.

Critical Threats to Atm Infrastructure and Communication Networks

The Air Traffic Management (ATM) infrastructure and communication networks are vital for the safe and efficient operation of air travel. However, they face various critical threats that can compromise their integrity, availability, and security. Understanding these threats is essential for developing effective risk management strategies and implementing robust cybersecurity measures. Here's an overview of the critical threats to ATM infrastructure and communication networks:

1. Cybersecurity Threats

a. Malware Attacks

- **Description**: Malicious software designed to disrupt, damage, or gain unauthorized access to computer systems can target ATM infrastructure.

- **Impact**: Malware can compromise the functionality of ATM systems, leading to operational failures, unauthorized access to sensitive data, and potential safety hazards.

b. Ransomware

- **Description**: Ransomware encrypts critical data or systems and demands payment for decryption keys.

- **Impact**: An effective ransomware attack on ATM systems could lead to widespread disruption, forcing air traffic control to revert to manual processes, significantly increasing the risk of errors and accidents.

c. Phishing Attacks

- **Description**: Cybercriminals may use phishing techniques to trick employees into divulging credentials or installing malicious software.

- **Impact**: Successful phishing attempts can provide attackers with access to critical systems, enabling further attacks or data breaches.

2. Insider Threats

a. Malicious Insiders

- **Description**: Employees or contractors with authorized access may intentionally misuse their privileges for personal gain or to cause disruption.
- **Impact**: Insider threats can lead to data breaches, sabotage of systems, or manipulation of traffic control operations, jeopardizing passenger safety.

b. Negligent Insiders

- **Description**: Unintentional actions by employees, such as mishandling sensitive information or failing to follow security protocols, can create vulnerabilities.
- **Impact**: Human error can lead to security breaches, data leaks, or operational disruptions, compromising the integrity of ATM systems.

3. Physical Threats

a. Terrorism and Sabotage

- **Description**: Physical attacks on ATM facilities, communication networks, or critical infrastructure can disrupt air traffic operations.
- **Impact**: Acts of terrorism or sabotage can cause extensive damage, leading to loss of life, financial losses, and long-term operational challenges.

b. Natural Disasters

- **Description**: Earthquakes, floods, hurricanes, or other natural events can physically damage ATM infrastructure and communication networks.

- **Impact**: Natural disasters can disrupt operations, damage critical equipment, and hinder communication, complicating air traffic management and emergency responses.

4. Technological Vulnerabilities

a. Outdated Systems

- **Description**: Legacy systems may lack modern security features and are often more vulnerable to attacks.

- **Impact**: Utilizing outdated technology increases the risk of exploitation by cybercriminals and can hinder the effectiveness of ATM operations.

b. Interconnected Systems

- **Description**: The reliance on interconnected systems and networks for information sharing can create vulnerabilities.

- **Impact**: A compromise in one system can lead to a cascading effect, impacting multiple systems and compromising overall ATM integrity.

5. Supply Chain Vulnerabilities

- **Description**: Third-party vendors and contractors involved in the provision of technology and services can introduce security risks.

- **Impact**: Vulnerabilities in the supply chain can lead to data breaches or operational failures, as attackers may exploit weak points in the relationships between ATM operators and their suppliers.

6. Communication Network Threats

a. Denial of Service (DoS) Attacks

- **Description**: Attackers may overwhelm communication networks with excessive traffic, rendering them inoperable.

- **Impact**: A successful DoS attack can disrupt communications between air traffic control and aircraft, leading to potential safety hazards and operational inefficiencies.

b. Eavesdropping and Signal Interception

- **Description**: Intercepting radio communications between air traffic controllers and pilots can provide attackers with sensitive operational information.

- **Impact**: Eavesdropping can compromise the confidentiality of communication, leading to unauthorized access to operational details and potential misuse.

7. Regulatory and Compliance Risks

- **Description**: Non-compliance with cybersecurity regulations and standards can lead to vulnerabilities and increased risks.

- **Impact**: Regulatory breaches can result in penalties, legal liabilities, and increased scrutiny from regulatory bodies, impacting operational integrity and public trust.

Securing air-to-ground communications is a critical aspect of air traffic management (ATM) and aviation safety. Effective communication between aircraft and ground-based air traffic control is essential for safe operations, and any vulnerabilities in this communication channel can have severe consequences. This overview outlines the key aspects of securing air-to-ground communications, including technologies involved, common threats, and strategies for enhancing security.

1. Overview of Air-to-Ground Communications

Air-to-ground communications are essential for:

- **Flight Operations**: Allowing pilots to receive instructions, weather updates, and information about air traffic and routes.

- **Safety Management**: Enabling rapid communication in emergencies or changes in flight plans.

- **Coordination**: Facilitating coordination between different aviation stakeholders, including air traffic control (ATC), airlines, and airport authorities.

2. Technologies Used in Air-to-Ground Communications

Air-to-ground communication relies on several technologies, including:

a. Radio Communication

- **VHF (Very High Frequency) Radio**: Primarily used for voice communication between pilots and air traffic controllers.

- **HF (High Frequency) Radio**: Used for long-range communications, particularly over oceanic routes.

b. Data Link Communication

- **Controller-Pilot Data Link Communications (CPDLC)**: Allows for text-based communication between pilots and air traffic controllers, reducing reliance on voice communication.

- **Automatic Dependent Surveillance-Broadcast (ADS-B)**: Provides real-time aircraft position and intent information, enhancing situational awareness.

c. Satellite Communication

- **Inmarsat and Iridium**: Satellite systems that provide voice and data communication for aircraft flying in remote areas where traditional ground-based systems are unavailable.

3. Common Threats to Air-to-Ground Communications

Air-to-ground communication systems face several threats, including:

a. Cyber Attacks

- **Interception of Communications**: Unauthorized access to communication channels can lead to eavesdropping or spoofing of messages.

- **Denial of Service (DoS)**: Overloading communication networks can disrupt service, hindering the ability of pilots and ATC to communicate effectively.

b. Physical Interference

- **Jamming**: Intentional interference with radio signals can prevent communication between aircraft and ground stations.

- **Signal Interception**: Adversaries can intercept communications to gather intelligence or disrupt operations.

c. Insider Threats

- **Negligent or Malicious Employees**: Personnel with access to communication systems may inadvertently or intentionally compromise security.

4. Strategies for Securing Air-to-Ground Communications

To enhance the security of air-to-ground communications, several strategies can be employed:

a. Encryption

- **Data Encryption**: Encrypting voice and data transmissions ensures that intercepted communications remain unreadable. This is crucial for protecting sensitive operational information.

- **End-to-End Encryption**: Implementing encryption between the cockpit and ground control enhances security against eavesdropping and tampering.

b. Authentication and Access Control

- **User Authentication**: Implementing strong authentication mechanisms (e.g., multi-factor authentication) for personnel accessing communication systems reduces the risk of unauthorized access.

- **Role-Based Access Control**: Ensuring that only authorized personnel can access specific systems and information helps mitigate insider threats.

c. Network Security Measures

- **Firewalls and Intrusion Detection Systems (IDS)**: Deploying robust firewalls and IDS to monitor and protect communication networks against unauthorized access and cyber threats.

- **Segmentation**: Isolating critical communication systems from other networks to reduce the risk of attacks.

d. Regular Security Assessments

- **Penetration Testing**: Conducting regular penetration tests and vulnerability assessments to identify and address potential weaknesses in air-to-ground communication systems.

- **Incident Response Planning**: Developing and testing incident response plans to ensure a coordinated approach to handling security breaches or disruptions.

e. Training and Awareness

- **Employee Training**: Providing training for all personnel on cybersecurity best practices, including recognizing phishing attempts and secure handling of sensitive information.

- **Simulations and Drills**: Conducting regular simulations and drills to prepare for potential security incidents and reinforce protocols.

f. Collaboration and Information Sharing

- **Industry Collaboration**: Working with industry partners, regulatory bodies, and government agencies to share information about emerging threats and best practices for securing communications.

- **Participating in Cybersecurity Initiatives**: Engaging in collaborative initiatives aimed at enhancing aviation cybersecurity across the industry.

Chapter 8

Data Protection and Privacy in Aviation

Importance Of Passenger and Crew Data Protection

The protection of passenger and crew data is of paramount importance in the aviation industry, driven by legal, ethical, operational, and reputational considerations. With the increasing reliance on digital systems and the collection of vast amounts of personal data, ensuring the security and privacy of this information is essential for maintaining trust, compliance, and overall operational integrity. Here are the key aspects highlighting the importance of protecting passenger and crew data:

1. Legal and Regulatory Compliance

- **Data Protection Laws**: Various jurisdictions have enacted stringent data protection laws, such as the General Data Protection Regulation (GDPR) in the European Union and the California Consumer Privacy Act (CCPA) in the United States. These regulations impose strict requirements on how organizations collect, store, process, and share personal data.

- **Compliance Penalties**: Failure to comply with data protection laws can result in severe financial penalties, legal repercussions, and enforcement actions by regulatory authorities, potentially jeopardizing an airline's operations and financial stability.

2. Protecting Sensitive Information

- **Personal Identifiable Information (PII)**: Passenger and crew data often include sensitive information such as names, addresses, passport numbers, and credit card details. Unauthorized access or data breaches can lead to identity theft, financial fraud, and other malicious activities.

- **Health Data**: In the context of medical emergencies or health-related data collection (e.g., health declarations during pandemics), protecting sensitive health information is crucial to maintaining patient confidentiality and trust.

3. Maintaining Customer Trust and Loyalty

- **Trust and Reputation**: Passengers expect their personal information to be handled securely and responsibly. A data breach or mishandling of personal data can significantly damage an airline's reputation and erode customer trust, leading to lost business and customer loyalty.

- **Transparency and Accountability**: Airlines that prioritize data protection and communicate their commitment to safeguarding personal information foster trust among passengers and crew, encouraging loyalty and positive brand perception.

4. Operational Integrity and Safety

- **Critical Operational Data**: Crew data, including qualifications, training records, and performance metrics, are essential for maintaining operational safety and compliance.

Protecting this information ensures that only qualified personnel operate aircraft and manage flight operations.

- **Risk Management**: Effective data protection measures contribute to overall risk management strategies, ensuring that operational processes remain secure and resilient against cyber threats.

5. Mitigating Cybersecurity Risks

- **Cyber Threat Landscape**: The aviation industry is increasingly targeted by cybercriminals seeking to exploit vulnerabilities in data systems. Protecting passenger and crew data is essential for mitigating risks associated with data breaches, ransomware attacks, and other cyber threats.

- **Incident Response Preparedness**: Establishing robust data protection measures and incident response protocols enhances an organization's ability to respond to data breaches swiftly and effectively, minimizing potential damage and operational disruptions.

6. Enhancing Data Management Practices

- **Data Minimization**: Implementing data protection measures encourages airlines to adopt data minimization practices, collecting only the information necessary for operational purposes. This reduces the risk of data exposure and enhances overall security.

- **Regular Audits and Assessments**: Ongoing evaluation of data protection practices promotes accountability and continuous improvement, ensuring that data management systems remain secure and compliant with evolving regulations.

7. Global Travel Context

- **International Regulations**: As airlines operate globally, they must navigate varying data protection regulations in different countries. Understanding and complying with these diverse requirements is crucial for protecting passenger and crew data across borders.

- **Cross-Border Data Transfers**: Safeguarding personal data during cross-border transfers is essential to comply with international data protection laws and maintain passenger trust in an interconnected global travel environment.

Compliance with data protection laws, such as the General Data Protection Regulation (GDPR) and the California Consumer Privacy Act (CCPA), is essential for airlines and aviation organizations that handle personal data of passengers and crew. These regulations impose stringent requirements on how organizations collect, process, store, and share personal information, and failure to comply can lead to significant penalties, reputational damage, and legal consequences. Here's an overview of the key aspects of compliance with these laws:

1. General Data Protection Regulation (GDPR)

The GDPR, enacted in May 2018, sets forth comprehensive regulations for the protection of personal data in the European Union (EU) and the European Economic Area (EEA). Key components of GDPR compliance include:

a. Scope and Applicability

- **Geographical Reach**: GDPR applies to any organization processing personal data of individuals located in the EU, regardless of where the organization is based. This means that airlines operating outside the EU must still comply when dealing with EU citizens' data.

- **Definition of Personal Data**: Personal data is defined broadly under GDPR and includes any information relating to an identified or identifiable natural person (e.g., names, email addresses, passport numbers).

b. Lawful Basis for Processing

- **Consent**: Organizations must obtain explicit consent from individuals before processing their personal data for specific purposes.

- **Contractual Necessity**: Processing is allowed when necessary for the performance of a contract (e.g., ticket purchases).

- **Legal Obligations**: Organizations may process data to comply with legal requirements (e.g., safety regulations).

c. Rights of Data Subjects

- **Right to Access**: Individuals have the right to request access to their personal data held by organizations.

- **Right to Rectification**: Individuals can request corrections to inaccurate personal data.

- **Right to Erasure**: Also known as the "right to be forgotten," individuals can request the deletion of their data under certain conditions.

- **Right to Data Portability**: Individuals can request their personal data in a structured, commonly used format for transfer to another organization.

d. Data Protection by Design and by Default

- Organizations must implement data protection measures at the design phase of new products or processes and ensure that only necessary data is processed by default.

e. Data Breach Notification

- Organizations are required to report personal data breaches to the relevant supervisory authority within 72 hours and, in some cases, notify affected individuals.

2. California Consumer Privacy Act (CCPA)

The CCPA, which took effect on January 1, 2020, is a state law that enhances privacy rights and consumer protection for residents of California. Key aspects of CCPA compliance include:

a. Scope and Applicability

- **Covered Businesses**: The CCPA applies to businesses that meet certain thresholds, including gross revenue over $25 million, buy or sell personal information of 50,000 or more consumers, or derive 50% or more of their annual revenues from selling personal information.

- **Definition of Personal Information**: Personal information under CCPA includes any data that identifies, relates to, or could reasonably be linked to a consumer or household.

b. Consumer Rights

- **Right to Know**: Consumers have the right to know what personal information is collected, used, shared, or sold by businesses.

- **Right to Delete**: Consumers can request the deletion of their personal data held by businesses, subject to certain exceptions.

- **Right to Opt-Out**: Consumers have the right to opt-out of the sale of their personal information to third parties.

c. Privacy Notices

- Businesses must provide clear and transparent privacy notices informing consumers about data collection practices, the purpose of data processing, and consumers' rights under the CCPA.

d. Non-Discrimination

- Businesses cannot discriminate against consumers who exercise their rights under the CCPA, such as by denying services or charging different prices.

3. Key Steps for Compliance

To comply with GDPR, CCPA, and other data protection laws, aviation organizations can take the following steps:

a. Data Inventory and Mapping

- Conduct a thorough inventory of personal data collected, processed, and stored, including data flow mapping to understand how data moves within and outside the organization.

b. Privacy Policies and Notices

- Develop and maintain clear and comprehensive privacy policies and notices that outline data collection practices, rights, and responsibilities in compliance with applicable laws.

c. Implement Data Protection Measures

- Establish strong data security measures, including encryption, access controls, and regular security assessments, to protect personal data from unauthorized access and breaches.

d. Training and Awareness

- Provide regular training to employees on data protection laws, organizational policies, and best practices for handling personal data.

e. Appoint a Data Protection Officer (DPO)

- For organizations subject to GDPR, appoint a DPO to oversee data protection compliance, act as a point of contact for data subjects and regulatory authorities, and ensure adherence to data protection principles.

f. Conduct Regular Audits and Assessments

- Implement regular audits to assess compliance with data protection laws, identify areas for improvement, and address any potential vulnerabilities in data handling practices.

Safeguarding sensitive information is crucial for aviation organizations, which handle vast amounts of personal and operational data. Implementing robust security strategies can protect this data from unauthorized access, breaches, and other threats. Here's an overview of effective strategies for safeguarding sensitive information:

1. Data Classification and Inventory

- **Identify and Classify Data**: Conduct a thorough inventory of all data assets and classify them based on sensitivity (e.g., public, confidential, sensitive personal data). This helps prioritize protection efforts for the most critical data.

- **Establish Data Handling Procedures**: Develop specific procedures for handling different classes of data, ensuring that sensitive information receives the highest level of protection.

2. Access Control Measures

- **Role-Based Access Control (RBAC)**: Implement RBAC to ensure that employees have access only to the data necessary for their roles. This minimizes exposure to sensitive information.

- **Least Privilege Principle**: Grant users the minimum level of access required to perform their job functions, reducing the risk of unauthorized access or accidental data exposure.

- **Strong Authentication**: Utilize strong authentication mechanisms, such as multi-factor authentication (MFA), to verify user identities before granting access to sensitive data.

3. Data Encryption

- **Encryption at Rest**: Encrypt sensitive data stored on servers, databases, and other storage devices to protect it from unauthorized access.

- **Encryption in Transit**: Use encryption protocols (e.g., TLS) to protect data transmitted over networks, safeguarding it from interception during transmission.

- **End-to-End Encryption**: For particularly sensitive communications, implement end-to-end encryption to ensure that data remains secure throughout its lifecycle, from sender to recipient.

4. Regular Security Assessments

- **Vulnerability Assessments**: Conduct regular vulnerability assessments and penetration testing to identify and remediate potential weaknesses in data systems and processes.

- **Security Audits**: Perform periodic security audits to evaluate compliance with data protection policies and regulations, ensuring that safeguards remain effective.

5. Data Minimization and Retention Policies

- **Data Minimization**: Collect and retain only the data necessary for specific business purposes. This reduces the risk of exposure and simplifies data management.

- **Retention Policies**: Establish clear data retention policies that define how long sensitive information will be kept and when it should be securely disposed of. Ensure compliance with relevant legal and regulatory requirements.

6. Incident Response Planning

- **Develop an Incident Response Plan**: Create a comprehensive incident response plan that outlines procedures for responding to data breaches or security incidents. This plan should include steps for containment, investigation, notification, and remediation.

- **Regular Testing**: Conduct regular simulations and drills to test the incident response plan, ensuring that staff are familiar with their roles and responsibilities in the event of a data breach.

7. Employee Training and Awareness

- **Ongoing Training**: Provide regular training on data protection best practices, cybersecurity awareness, and the importance of safeguarding sensitive information. Ensure that employees understand their roles in maintaining data security.

- **Phishing Simulations**: Conduct phishing simulations to educate employees about recognizing and responding to phishing attempts and other social engineering attacks.

8. Secure Third-Party Relationships

- **Vendor Risk Management**: Assess the data protection practices of third-party vendors and partners who handle sensitive information on behalf of the organization. Ensure that they adhere to similar security standards.

- **Data Processing Agreements**: Establish data processing agreements (DPAs) with third parties that outline responsibilities for data protection, breach notification, and compliance with relevant laws.

9. Physical Security Measures

- **Access Controls**: Implement physical access controls (e.g., keycards, biometric scanners) to secure areas where sensitive data is stored or processed, such as data centers and server rooms.

- **Surveillance and Monitoring**: Use surveillance cameras and monitoring systems to detect unauthorized access or suspicious activity in sensitive areas.

10. Data Loss Prevention (DLP) Solutions

- **DLP Technologies**: Implement DLP solutions to monitor, detect, and prevent the unauthorized transfer or sharing of sensitive information, both in transit and at rest.

- **Automated Alerts**: Set up automated alerts for potential data breaches or policy violations, enabling timely responses to security incidents.

11. Regular Policy Review and Updates

- **Data Protection Policies**: Regularly review and update data protection policies to reflect changes in regulations, emerging threats, and organizational practices.

- **Stakeholder Involvement**: Involve relevant stakeholders in policy development and updates to ensure that all perspectives are considered and that policies remain practical and effective.

12. Secure Cloud Practices

- **Cloud Security Measures**: When using cloud services, ensure that appropriate security measures are in place, including encryption, access controls, and compliance with data protection standards.

- **Vendor Security Assessment**: Evaluate the security practices of cloud service providers to ensure they align with organizational security requirements and regulatory obligations.

13. Collaboration and Information Sharing

- **Industry Collaboration**: Participate in industry forums and information-sharing initiatives to stay informed about emerging threats and best practices for data protection.

- **Government Partnerships**: Collaborate with government agencies and cybersecurity organizations to enhance threat intelligence and strengthen overall security posture.

Chapter 9

Industry Regulations and Standards

Overview Of Cybersecurity Regulations in Aviation

Cybersecurity regulations in the aviation sector are crucial for protecting sensitive data and ensuring the safety and security of air travel. Various international and national organizations have developed frameworks and guidelines to address cybersecurity risks. Here's an overview of key regulatory bodies and their respective regulations in aviation cybersecurity:

1. International Civil Aviation Organization (ICAO)

- **Role**: ICAO is a specialized agency of the United Nations that establishes global standards and regulations for civil aviation safety, security, and environmental protection.

- **Cybersecurity Framework**: In 2021, ICAO adopted the *Cybersecurity Framework* to provide guidance for member states on managing cybersecurity risks in civil aviation. This framework emphasizes a risk-based approach to cybersecurity and includes the following components:

  - **Cybersecurity Risk Management**: States are encouraged to implement risk management processes that include identifying, assessing, and mitigating cybersecurity risks associated with aviation systems.

  - **National Cybersecurity Policies**: Member states should develop and implement national policies and regulatory frameworks that incorporate cybersecurity measures for the aviation sector.

- o **Collaboration and Information Sharing**: ICAO promotes collaboration among stakeholders, including government agencies, industry players, and international organizations, to enhance cybersecurity preparedness and resilience.

- **Annex 17 - Security**: ICAO's Annex 17 to the Convention on International Civil Aviation provides standards and recommended practices (SARPs) for aviation security, including provisions related to cybersecurity. States are required to ensure that aviation security measures address potential cyber threats to aviation systems and operations.

2. European Union Aviation Safety Agency (EASA)

- **Role**: EASA is responsible for promoting the highest standards of safety and environmental protection in civil aviation across Europe. It provides regulatory oversight and guidance to member states and aviation stakeholders.

- **Regulation (EU) 2019/1383**: This regulation, effective from 2020, establishes a framework for aviation cybersecurity in Europe. Key elements include:

  - o **Security Management System (SeMS)**: Organizations are required to implement a SeMS that identifies cybersecurity risks, assesses vulnerabilities, and defines mitigation strategies.

  - o **Reporting Obligations**: Aviation organizations must report significant cybersecurity incidents to EASA, fostering a culture of transparency and collaboration in addressing cyber threats.

  - o **Continuous Improvement**: EASA emphasizes the need for continuous monitoring, evaluation, and improvement of cybersecurity measures within aviation organizations.

- **Guidelines for Cybersecurity in Aviation**: EASA has published various guidelines and materials to assist aviation stakeholders in implementing effective cybersecurity measures. These guidelines provide best practices for addressing cybersecurity risks across different areas, including aircraft, airports, and air traffic management.

3. Federal Aviation Administration (FAA)

- **Role**: The FAA is the regulatory body for civil aviation in the United States. It oversees the safety and security of the national airspace system and works to promote aviation cybersecurity.

- **Cybersecurity Policy**: The FAA's *Cybersecurity Strategy* focuses on protecting critical infrastructure within the aviation ecosystem. Key components include:

  - **Risk Management**: The FAA employs a risk-based approach to identify and mitigate cybersecurity risks to aviation operations and systems.

  - **Collaboration with Industry**: The FAA collaborates with industry partners, government agencies, and international organizations to enhance cybersecurity awareness and develop effective solutions to cyber threats.

- **Advisory Circulars**: The FAA issues advisory circulars (ACs) that provide guidance to aviation stakeholders on cybersecurity practices. For example:

  - **AC 120-112**: This circular provides guidance on cybersecurity measures for commercial operators and addresses risk management, incident reporting, and security management.

- **Security Directive**: In 2021, the FAA issued a security directive to address cybersecurity risks in pipeline transportation systems, which can also be applied to aviation systems. This directive requires operators to develop and implement plans to protect their systems from cyber threats.

4. National Institute of Standards and Technology (NIST)

- **Role**: NIST is a U.S. federal agency that develops standards, guidelines, and best practices for cybersecurity across various sectors, including aviation.

- **NIST Cybersecurity Framework**: NIST's Cybersecurity Framework (CSF) provides a flexible, risk-based approach to managing cybersecurity risks. Although not aviation-specific, the framework is widely adopted in the industry and can be tailored to meet the unique needs of aviation organizations.

- **Special Publications**: NIST also publishes specialized guidelines relevant to aviation cybersecurity, such as SP 800-53, which provides security controls for federal information systems and organizations.

5. Other Relevant Regulations and Guidelines

- **Federal Aviation Administration Modernization and Reform Act of 2012**: This act includes provisions for improving aviation safety and security, including addressing cybersecurity challenges.

- **Aviation Cybersecurity Action Plan**: The U.S. Department of Transportation (DOT) has developed an action plan to strengthen cybersecurity across transportation sectors, including aviation. The plan focuses on enhancing collaboration, information sharing, and incident response capabilities.

- **International Air Transport Association (IATA) Guidelines**: IATA provides guidelines and best practices for airlines and aviation stakeholders to enhance cybersecurity measures and resilience against cyber threats.

Standards and guidelines for cybersecurity in aviation provide frameworks and best practices that organizations can adopt to protect their systems, data, and operations from cyber threats. Various international, regional, and national organizations have developed these standards to address the unique challenges faced by the aviation industry. Here's an overview of key standards and guidelines relevant to cybersecurity in aviation:

1. International Civil Aviation Organization (ICAO)

- **ICAO Cybersecurity Framework**: Adopted in 2021, this framework outlines a comprehensive approach for states to manage cybersecurity risks in civil aviation. It includes guidance on:

    o   Establishing cybersecurity policies and regulations.

    o   Implementing a risk management approach to identify and mitigate threats.

    o   Collaborating and sharing information among stakeholders.

- **Annex 17 - Security**: This annex of the Convention on International Civil Aviation sets out international standards and recommended practices (SARPs) for aviation security, including the integration of cybersecurity measures into aviation security frameworks.

- **Guidance Material**: ICAO provides various guidance materials and documents to assist member states and stakeholders in implementing effective cybersecurity practices, including risk assessment methodologies and incident response plans.

2. European Union Aviation Safety Agency (EASA)

- **Regulation (EU) 2019/1383**: This regulation establishes a framework for aviation cybersecurity within the European Union. Key components include:

  - Requirement for a Security Management System (SeMS) for aviation organizations.

  - Incident reporting obligations to EASA.

  - Guidelines for assessing and managing cybersecurity risks.

- **EASA Guidelines for Cybersecurity in Aviation**: EASA has published guidance material that offers best practices for managing cybersecurity risks across different domains, including aircraft, airports, and air traffic management.

3. Federal Aviation Administration (FAA)

- **Advisory Circulars (AC)**: The FAA issues ACs that provide recommendations and guidelines for various aspects of aviation safety and security. Key circulars related to cybersecurity include:

  - **AC 120-112**: Guidance for commercial operators on cybersecurity measures, including risk management and incident reporting.

- **Cybersecurity Strategy**: The FAA's strategy emphasizes collaboration with industry stakeholders, risk management, and the implementation of cybersecurity measures tailored to the aviation sector.

4. National Institute of Standards and Technology (NIST)

- **NIST Cybersecurity Framework (CSF)**: Although not aviation-specific, the CSF is widely used across various industries, including aviation, to manage cybersecurity risks.

The framework consists of five core functions: Identify, Protect, Detect, Respond, and Recover.

- **NIST Special Publications**: NIST has developed several publications that provide guidelines for securing information systems, including:

    o **SP 800-53**: Security and Privacy Controls for Information Systems and Organizations, which includes a comprehensive set of controls that can be applied to aviation systems.

    o **SP 800-171**: Protecting Controlled Unclassified Information in Nonfederal Systems and Organizations, relevant for aviation contractors handling sensitive data.

5. International Air Transport Association (IATA)

- **IATA Cybersecurity Framework**: IATA has developed a framework and guidelines for airlines to enhance their cybersecurity posture. The framework includes:

    o Risk assessment and management practices.

    o Recommendations for incident response and recovery.

    o Best practices for securing data and protecting passenger information.

- **Guidelines for Information Security Management Systems**: IATA provides guidance for developing and implementing information security management systems (ISMS) tailored to the aviation industry.

6. ISO/IEC 27000 Series

- **ISO/IEC 27001**: This international standard provides requirements for establishing, implementing, maintaining, and continuously improving an information security management system (ISMS). It is applicable to all types of organizations, including those in the aviation sector.

- **ISO/IEC 27002**: This standard offers guidelines for organizational information security practices, including the selection and implementation of security controls, making it a useful reference for aviation organizations.

7. Cybersecurity and Infrastructure Security Agency (CISA)

- **CISA Cybersecurity Framework**: CISA provides resources and guidelines for enhancing cybersecurity across critical infrastructure sectors, including aviation. Their resources include:

  o Best practices for securing industrial control systems (ICS) and operational technology (OT).

  o Guidelines for incident reporting and sharing information about cyber threats.

8. Other Relevant Standards and Guidelines

- **NATO Cybersecurity Framework**: NATO has developed cybersecurity standards and guidelines that can be relevant to aviation organizations engaged in defense and military operations.

- **SANS Institute Guidelines**: The SANS Institute provides numerous resources, including whitepapers and training programs focused on cybersecurity best practices, many of which can be applied to the aviation sector.

Emerging policies and international cooperation play a vital role in strengthening cybersecurity in aviation. As the aviation sector continues to evolve with advancements in technology, the associated cybersecurity risks grow more complex and interconnected. In response, governments, organizations, and industry stakeholders are increasingly recognizing the need for cohesive strategies and collaborative efforts. Here's an overview of the emerging policies and the role of international cooperation in enhancing aviation cybersecurity:

1. Emerging Policies in Aviation Cybersecurity

a. National Cybersecurity Strategies

Many countries are developing or updating national cybersecurity strategies that specifically address the aviation sector. These strategies typically include:

- **Risk Assessment**: Frameworks for identifying and assessing cybersecurity risks specific to aviation infrastructure and operations.

- **Cybersecurity Frameworks**: Adoption of established frameworks, such as the NIST Cybersecurity Framework, tailored for aviation stakeholders.

- **Incident Response Plans**: Development of protocols for responding to cybersecurity incidents, including communication strategies and recovery plans.

- **Training and Awareness Programs**: Initiatives to enhance cybersecurity awareness and training for aviation personnel, including pilots, ground crew, and IT staff.

b. Regulatory Frameworks

Regulatory bodies are increasingly enacting legislation and regulations focused on cybersecurity in aviation, including:

- **Mandatory Reporting of Cyber Incidents**: Regulations requiring airlines, airports, and other aviation stakeholders to report significant cybersecurity incidents to relevant authorities.

- **Compliance Standards**: Establishing mandatory compliance standards for cybersecurity practices, such as those outlined by EASA and the FAA.

- **Periodic Audits and Assessments**: Mandating regular audits and assessments of cybersecurity measures in aviation organizations to ensure compliance and identify vulnerabilities.

c. Data Protection and Privacy Regulations

With the rise of data breaches and privacy concerns, many countries are implementing stringent data protection regulations that affect the aviation sector. Key elements include:

- **Personal Data Protection Laws**: Regulations like the General Data Protection Regulation (GDPR) in Europe and the California Consumer Privacy Act (CCPA) in the U.S. that impose strict requirements on how organizations collect, store, and protect personal data.

- **Cross-Border Data Transfer**: Policies governing the transfer of personal data across borders, requiring organizations to ensure that cybersecurity measures are in place when transferring data internationally.

2. Role of International Cooperation

a. Global Standards Development

International cooperation is essential for developing and harmonizing cybersecurity standards and guidelines across countries. Organizations such as ICAO and EASA work collaboratively with member states to create:

- **Global Cybersecurity Frameworks**: Establishing standards that member states can adopt to enhance their cybersecurity posture while ensuring a unified approach to aviation security.

- **Best Practices Sharing**: Facilitating the exchange of information on cybersecurity threats, vulnerabilities, and mitigation strategies among countries and industry stakeholders.

b. Joint Exercises and Training

International cooperation fosters joint cybersecurity exercises and training programs that enhance preparedness and response capabilities among aviation stakeholders:

- **Simulated Cyber Incidents**: Conducting joint exercises that simulate cyber incidents to test and improve the incident response capabilities of participating organizations.

- **Cross-Training Programs**: Developing training programs that allow aviation personnel from different countries to learn from each other's experiences and best practices in cybersecurity.

c. Information Sharing and Threat Intelligence

Collaboration among nations and industry partners enhances information sharing on emerging threats and vulnerabilities:

- **Threat Intelligence Platforms**: Establishing platforms for sharing threat intelligence, allowing aviation organizations to stay informed about the latest cyber threats and vulnerabilities.

- **Public-Private Partnerships**: Encouraging collaboration between government agencies and private sector organizations to enhance information sharing and improve overall cybersecurity resilience in the aviation sector.

d. International Agreements and Initiatives

Various international agreements and initiatives aim to strengthen cooperation in aviation cybersecurity:

- **Bilateral and Multilateral Agreements**: Countries are forming agreements to enhance cooperation in cybersecurity, share best practices, and conduct joint training exercises.

- **Global Initiatives**: Participation in global initiatives, such as the United Nations' Global Cybersecurity Agenda, which promotes international cooperation on cybersecurity issues across various sectors, including aviation.

3. Challenges to International Cooperation

While emerging policies and international cooperation are crucial for enhancing aviation cybersecurity, challenges remain:

- **Diverse Regulatory Environments**: Differences in regulatory frameworks and standards across countries can hinder cooperation and complicate compliance for international aviation organizations.

- **Resource Limitations**: Many countries, particularly developing nations, may lack the resources and expertise to implement robust cybersecurity measures and participate in international cooperation efforts.

- **Rapidly Evolving Threat Landscape**: The fast-paced nature of cyber threats necessitates continuous updates and adaptations to policies and cooperation frameworks, which can be challenging to maintain.

Chapter 10

Cybersecurity Tools and Technologies

Key Cybersecurity Technologies

Key cybersecurity technologies are essential for protecting aviation systems and data from cyber threats. These technologies play critical roles in safeguarding sensitive information, ensuring system integrity, and enabling organizations to respond effectively to potential breaches. Here's an overview of three fundamental cybersecurity technologies: encryption, firewalls, and intrusion detection systems (IDS).

1. Encryption

Overview: Encryption is a method of converting plaintext data into an unreadable format (ciphertext) using algorithms and keys. This process ensures that only authorized users with the correct decryption key can access the original information.

Importance in Aviation:

- **Data Protection**: Encryption protects sensitive data transmitted over networks (such as passenger information and flight data) from unauthorized access, ensuring confidentiality and integrity.

- **Regulatory Compliance**: Many data protection regulations, like GDPR and CCPA, require organizations to implement encryption as a safeguard for personal data.

- **Secure Communication**: Encryption is critical for securing communications between various systems in aviation, including air-to-ground communications, airport operations, and data shared among stakeholders.

Types of Encryption:

- **Symmetric Encryption**: Uses the same key for both encryption and decryption. It is faster but requires secure key management.

- **Asymmetric Encryption**: Utilizes a pair of keys (public and private) for encryption and decryption. It enhances security, especially for data exchanged over insecure networks.

2. Firewalls

Overview: Firewalls are network security devices or software applications that monitor and control incoming and outgoing network traffic based on predetermined security rules. They act as a barrier between trusted internal networks and untrusted external networks.

Importance in Aviation:

- **Access Control**: Firewalls enforce access controls by allowing or blocking traffic based on security policies, protecting sensitive aviation systems from unauthorized access.

- **Threat Prevention**: They help prevent malware, ransomware, and other malicious activities from entering the network, thereby reducing the risk of cyber incidents.

- **Network Segmentation**: Firewalls enable the segmentation of networks, isolating critical systems (e.g., air traffic control and flight operations) from less secure areas, which limits potential damage in case of a breach.

Types of Firewalls:

- **Packet-Filtering Firewalls**: Analyze packets of data and accept or reject them based on established rules.

- **Stateful Inspection Firewalls**: Monitor the state of active connections and make decisions based on the context of the traffic.

- **Next-Generation Firewalls (NGFW)**: Combine traditional firewall capabilities with advanced features such as application awareness, intrusion prevention, and deep packet inspection.

3. Intrusion Detection Systems (IDS)

Overview: Intrusion Detection Systems are security solutions that monitor network traffic for suspicious activity and potential threats. They analyze traffic patterns and compare them against known attack signatures or anomalous behavior to detect unauthorized access or policy violations.

Importance in Aviation:

- **Threat Detection**: IDS plays a crucial role in identifying potential security breaches or attacks on aviation systems in real time, enabling rapid response to incidents.

- **Incident Response**: By providing alerts and detailed information about detected threats, IDS assists security teams in investigating and mitigating attacks effectively.

- **Compliance and Reporting**: IDS can generate logs and reports necessary for compliance with regulatory requirements, providing evidence of security posture and incident response actions.

Types of IDS:

- **Network Intrusion Detection Systems (NIDS)**: Monitor network traffic and analyze it for signs of intrusion or abnormal behavior.

- **Host Intrusion Detection Systems (HIDS)**: Monitor individual hosts or devices for unauthorized activity and policy violations.

- **Hybrid IDS**: Combine features of both NIDS and HIDS, offering a comprehensive view of security across the network and hosts.

Artificial intelligence (AI) and machine learning (ML) play a transformative role in enhancing threat detection and cybersecurity measures within the aviation industry. These technologies leverage vast amounts of data and sophisticated algorithms to identify patterns, detect anomalies, and respond to emerging cyber threats more efficiently than traditional methods. Here's an overview of the role of AI and ML in threat detection:

1. Enhanced Threat Detection

a. Anomaly Detection

- **Behavioral Analysis**: AI and ML systems can establish baseline behavior for normal network and user activity. By analyzing real-time data, these systems can identify deviations from typical patterns, which may indicate potential security breaches or intrusions.

- **Automated Monitoring**: Continuous monitoring of network traffic and user behavior enables faster detection of anomalies, allowing security teams to respond to threats before they escalate.

b. Signature-Based Detection

- **Pattern Recognition**: AI and ML can analyze historical data to recognize known attack signatures, facilitating the rapid identification of threats based on previously observed malicious behaviors.

- **Dynamic Signature Creation**: Rather than relying solely on static signatures, AI-driven systems can generate new signatures based on emerging threats, ensuring more effective defense against novel attack methods.

2. Improved Response Capabilities

a. Automated Incident Response

- **Real-Time Actions**: AI systems can automatically trigger predefined responses to detected threats, such as isolating compromised systems, blocking malicious traffic, or alerting security personnel. This reduces response times and minimizes damage during incidents.

- **Intelligent Decision-Making**: AI algorithms can analyze the severity and context of a threat, allowing for more nuanced responses that prioritize critical assets and vulnerabilities.

b. Predictive Analysis

- **Threat Forecasting**: By analyzing trends and patterns in cybersecurity incidents, AI can predict potential future threats, enabling organizations to take proactive measures to mitigate risks before they materialize.

- **Risk Assessment**: AI models can evaluate the risk associated with various vulnerabilities and recommend prioritization for security efforts, helping organizations allocate resources effectively.

3. Enhanced Security Operations

a. Security Information and Event Management (SIEM)

- **Data Correlation**: AI-enhanced SIEM systems can correlate data from various sources (e.g., firewalls, intrusion detection systems, and endpoint protection) to provide a comprehensive view of potential threats.

- **Intelligent Alerts**: AI can filter and prioritize alerts, reducing the noise generated by false positives and allowing security analysts to focus on the most critical incidents.

b. User and Entity Behavior Analytics (UEBA)

- **Contextual Awareness**: AI-driven UEBA solutions analyze user behavior and system interactions to identify unusual activities, such as unauthorized access attempts or data exfiltration, providing early warning signs of insider threats or compromised accounts.

- **Adaptive Learning**: Machine learning models can adapt to changes in user behavior over time, improving their accuracy in detecting anomalies.

4. Collaboration and Information Sharing

a. Threat Intelligence Platforms

- **Aggregating Threat Data**: AI can analyze data from various threat intelligence sources to identify emerging trends, vulnerabilities, and attack vectors relevant to the aviation industry.

- **Collaborative Defense**: AI-driven systems can facilitate information sharing between organizations, helping to build a collective defense against shared threats and improve overall situational awareness.

b. Automated Reporting

- **Incident Documentation**: AI can automate the documentation and reporting of cybersecurity incidents, streamlining the process and ensuring that critical information is captured for future analysis and compliance purposes.

5. Challenges and Considerations

While AI and ML offer significant advantages in threat detection, there are also challenges to consider:

- **Data Quality**: The effectiveness of AI and ML systems is heavily dependent on the quality and relevance of the data they analyze. Inaccurate or incomplete data can lead to false positives or missed threats.

- **Adversarial AI**: Cybercriminals may employ AI techniques to develop more sophisticated attacks, necessitating continuous adaptation and improvement of defensive measures.

- **Implementation Costs**: Integrating AI and ML into existing cybersecurity infrastructure can require significant investment in technology, talent, and training.

Advances in quantum computing represent a significant technological leap that has the potential to revolutionize various industries, including aviation. While the field is still in its developmental stages, the implications of quantum computing for aviation security are profound. This overview discusses the advancements in quantum computing and their potential impacts on aviation security, particularly concerning encryption, data protection, and overall cybersecurity.

1. Understanding Quantum Computing

a. Fundamentals of Quantum Computing

- **Quantum Mechanics Principles**: Quantum computing relies on principles of quantum mechanics, utilizing quantum bits (qubits) that can exist in multiple states simultaneously (superposition) and can be entangled, allowing for complex computations at unprecedented speeds.

- **Computational Power**: Quantum computers can solve certain types of problems much faster than classical computers, potentially outperforming them in tasks related to optimization, simulation, and cryptography.

2. Impact on Encryption and Data Security

a. Vulnerability of Classical Encryption

- **Breaking Traditional Cryptography**: Many of the encryption algorithms currently used to secure aviation data (such as RSA and ECC) could be rendered obsolete by quantum computers. Shor's algorithm, for instance, can factor large numbers exponentially faster than classical algorithms, posing a threat to public-key cryptography.

- **Implications for Aviation Security**: As aviation relies heavily on secure communication and data protection, the potential for quantum computers to break existing encryption poses significant risks, including unauthorized access to sensitive data such as passenger information, flight plans, and maintenance records.

b. Post-Quantum Cryptography

- **Development of Quantum-Resistant Algorithms**: To counter the threats posed by quantum computing, researchers are working on post-quantum cryptography (PQC) algorithms designed to be secure against quantum attacks. These algorithms will use mathematical problems that remain difficult for quantum computers to solve.

- **Implementation Challenges**: Transitioning to PQC in aviation systems will require significant investment in time and resources. Aviation organizations will need to update their infrastructure, train personnel, and ensure compatibility with existing systems.

3. Enhancements in Security Protocols

a. Quantum Key Distribution (QKD)

- **Secure Communication**: QKD uses the principles of quantum mechanics to create secure communication channels. It allows two parties to share encryption keys with absolute security, as any attempt at eavesdropping can be detected through changes in the quantum state.

- **Applications in Aviation**: QKD can enhance the security of communications between aircraft and ground control, as well as within airport systems, ensuring that sensitive data remains protected from interception or tampering.

b. Improved Authentication Mechanisms

- **Quantum-Based Authentication**: Quantum computing can enable more secure authentication mechanisms that leverage quantum properties to verify identities and access rights, reducing the risk of credential theft or unauthorized access.

4. Operational Improvements Through Quantum Computing

a. Optimization of Flight Operations

- **Route Optimization**: Quantum computing has the potential to optimize flight routes, taking into account numerous variables (weather, air traffic, fuel consumption) much more efficiently than classical methods.

- **Resource Management**: Enhanced data analysis capabilities can lead to improved management of airport resources, including baggage handling and scheduling, contributing to operational efficiency and safety.

b. Data Analysis for Threat Detection

- **Enhanced Threat Modeling**: Quantum computing can process vast amounts of data quickly, improving the ability to model and predict cyber threats. This capability can lead to more effective identification of vulnerabilities and potential attacks in real-time.

- **Simulation of Cyber Attacks**: Quantum computing can simulate complex cyber attack scenarios, allowing aviation security teams to better prepare for and respond to potential threats.

5. Challenges and Considerations

While the advancements in quantum computing hold promise, several challenges need to be addressed:

- **Development Timeline**: Quantum computing is still in the experimental phase, and practical applications in aviation security may take time to realize. Organizations must remain vigilant and proactive in their cybersecurity efforts in the meantime.

- **Investment and Resources**: Implementing quantum-resistant technologies and QKD systems will require substantial investment, both in technology and in the training of personnel.

- **Standardization and Regulation**: The aviation industry will need to establish standards and regulations regarding the adoption of quantum technologies, ensuring interoperability and security across international borders.

Chapter 11

Best Practices in Aviation Cybersecurity

Cyber Hygiene Practices for Airlines, Airports, And Passengers

Cyber hygiene practices are essential for airlines, airports, and passengers to ensure a secure and resilient aviation environment. By adopting a proactive approach to cybersecurity, stakeholders can mitigate risks, protect sensitive information, and enhance overall safety. Here's an overview of recommended cyber hygiene practices for each group:

1. Cyber Hygiene Practices for Airlines

a. Employee Training and Awareness

- **Regular Training Programs**: Conduct ongoing cybersecurity training for all employees, emphasizing the importance of recognizing phishing attempts, social engineering tactics, and other common cyber threats.

- **Simulated Phishing Exercises**: Implement regular phishing simulation tests to help employees identify and respond to suspicious emails and communications.

b. Access Control and Authentication

- **Role-Based Access Control (RBAC)**: Ensure that employees have access only to the information necessary for their roles. Implement the principle of least privilege to minimize risks.

- **Multi-Factor Authentication (MFA)**: Require MFA for all critical systems and applications, adding an extra layer of security to user accounts.

c. Incident Response Planning

- **Develop and Test an Incident Response Plan**: Create a comprehensive incident response plan that outlines roles, responsibilities, and procedures for handling cyber incidents. Regularly test and update the plan to ensure its effectiveness.

- **Establish a Cybersecurity Team**: Form a dedicated cybersecurity team responsible for monitoring threats, responding to incidents, and continuously improving security measures.

d. System Updates and Patch Management

- **Regular Software Updates**: Ensure that all systems, applications, and hardware are regularly updated to protect against known vulnerabilities.

- **Automated Patch Management**: Implement automated tools to manage and deploy patches promptly, reducing the window of exposure to threats.

e. Data Protection Measures

- **Data Encryption**: Encrypt sensitive data both in transit and at rest to protect it from unauthorized access.

- **Backup Procedures**: Establish regular backup procedures for critical data and systems to facilitate recovery in case of a cyber incident.

2. Cyber Hygiene Practices for Airports

a. Physical and Cyber Security Integration

- **Unified Security Operations**: Integrate physical security measures with cybersecurity practices to provide a comprehensive security posture. This includes monitoring access points and securing networks that support critical airport operations.

- **Surveillance and Monitoring**: Implement surveillance systems to monitor both physical and digital access to critical infrastructure and data centers.

b. Network Segmentation

- **Segment Critical Systems**: Isolate sensitive systems, such as air traffic control and passenger data systems, from less secure networks to minimize the impact of potential breaches.

- **Monitor Network Traffic**: Employ intrusion detection systems (IDS) and security information and event management (SIEM) solutions to monitor network traffic for suspicious activity.

c. Vendor and Third-Party Management

- **Assess Third-Party Risks**: Conduct thorough security assessments of vendors and third-party service providers to ensure they meet cybersecurity standards and practices.

- **Regular Audits and Compliance Checks**: Perform regular audits of third-party systems and services to ensure compliance with security policies.

d. Emergency Preparedness and Communication

- **Crisis Communication Plan**: Develop a crisis communication plan that includes protocols for informing passengers, employees, and stakeholders in the event of a cyber incident.

- **Coordination with Authorities**: Establish communication channels with local law enforcement and cybersecurity agencies to facilitate a coordinated response to cyber threats.

3. Cyber Hygiene Practices for Passengers

a. Personal Device Security

- **Use Strong Passwords**: Encourage passengers to use strong, unique passwords for their devices and accounts, and to enable MFA wherever possible.

- **Regular Software Updates**: Advise passengers to keep their devices and apps updated to protect against security vulnerabilities.

b. Secure Internet Practices

- **Avoid Public Wi-Fi for Sensitive Transactions**: Warn passengers against using public Wi-Fi networks for sensitive transactions, such as online banking or accessing personal accounts.

- **Use a Virtual Private Network (VPN)**: Encourage the use of VPNs when accessing public networks to encrypt internet traffic and protect personal information.

c. Awareness of Phishing and Scams

- **Educate on Phishing Risks**: Inform passengers about the risks of phishing emails and scams, and advise them to verify the authenticity of communications from airlines and travel agencies.

- **Report Suspicious Activities**: Encourage passengers to report any suspicious activities or communications to airline staff or cybersecurity authorities.

d. Data Protection

- **Limit Personal Information Sharing**: Advise passengers to limit the sharing of personal information, especially on social media or unsecured platforms.

- **Monitor Financial Accounts**: Encourage passengers to regularly monitor their bank and credit card statements for any unauthorized transactions.

Cybersecurity training for aviation personnel is critical to ensuring the safety and security of the aviation industry. As technology advances and cyber threats evolve, the need for a well-informed workforce becomes increasingly important. Here are several key reasons highlighting the importance of cybersecurity training for aviation personnel:

1. Raising Awareness of Cyber Threats

- **Understanding Risks**: Training helps personnel recognize the various cyber threats that can affect aviation operations, including phishing attacks, malware, ransomware, and social engineering tactics.

- **Identifying Vulnerabilities**: Personnel learn to identify potential vulnerabilities within their systems and processes, fostering a proactive approach to cybersecurity.

2. Preventing Human Error

- **Reducing Risk of Mistakes**: Human error is a significant factor in many cyber incidents. Training programs teach best practices and reinforce proper procedures to minimize the likelihood of errors that could lead to security breaches.

- **Promoting Vigilance**: Training instills a culture of vigilance, encouraging employees to be cautious and aware of their online activities and the potential risks associated with them.

3. Ensuring Compliance with Regulations

- **Meeting Regulatory Standards**: Many aviation regulatory bodies, such as the International Civil Aviation Organization (ICAO) and the Federal Aviation Administration (FAA), mandate cybersecurity training for personnel to ensure compliance with established standards.

- **Promoting Accountability**: Training emphasizes the importance of cybersecurity responsibilities, helping personnel understand their roles in maintaining a secure environment.

4. Enhancing Incident Response Capabilities

- **Preparedness for Cyber Incidents**: Training equips personnel with the knowledge and skills needed to respond effectively to cybersecurity incidents, reducing response times and mitigating the impact of breaches.

- **Simulation Exercises**: Conducting drills and simulations as part of training helps personnel practice their response to various scenarios, ensuring they are well-prepared to handle real incidents.

5. Fostering a Cybersecurity Culture

- **Creating a Security-Conscious Workforce**: Training promotes a culture of cybersecurity awareness throughout the organization, where every employee understands the importance of protecting sensitive information and systems.

- **Encouraging Reporting**: A strong cybersecurity culture encourages employees to report suspicious activities or potential threats without fear of reprisal, contributing to a more secure environment.

6. Supporting Technological Adaptation

- **Keeping Up with Advances**: As aviation technology evolves, so do the associated cyber threats. Training ensures personnel stay informed about the latest technologies and the security measures necessary to protect them.

- **Adapting to New Systems**: Employees need training when new systems or software are implemented to understand how to use them securely and recognize potential security implications.

7. Protecting Sensitive Information

- **Safeguarding Passenger Data**: Aviation personnel often handle sensitive information, including passenger data and financial information. Training helps them understand the importance of protecting this data from unauthorized access and breaches.

- **Maintaining Operational Integrity**: Effective cybersecurity training helps prevent data breaches that could disrupt operations, ensuring the integrity and reliability of aviation systems.

8. Encouraging Collaboration and Communication

- **Teamwork in Cybersecurity**: Training fosters collaboration among different departments and personnel levels, emphasizing the importance of working together to address cybersecurity challenges.

- **Building Communication Channels**: A trained workforce is better equipped to communicate effectively about cybersecurity issues, enabling rapid sharing of information about threats and vulnerabilities.

Building a resilient cybersecurity culture is essential for the aviation industry to protect against ever-evolving cyber threats and ensure operational safety and data security. A resilient cybersecurity culture goes beyond compliance, embedding cybersecurity principles into everyday practices and fostering a mindset of shared responsibility. Here are some effective strategies to develop a strong, enduring cybersecurity culture within an aviation organization:

1. Establish Strong Leadership and Commitment

- **Executive Support**: Leaders should actively endorse cybersecurity as a priority, demonstrating commitment by allocating resources and time to cybersecurity initiatives.

- **Cybersecurity Champions**: Designate cybersecurity champions within different departments to promote security awareness, facilitate communication, and provide guidance to team members.

2. Incorporate Cybersecurity into Organizational Values

- **Integrate into Core Values**: Position cybersecurity as a foundational value within the organization, emphasizing its role in protecting passenger safety, operational integrity, and business reputation.

- **Clear Communication of Importance**: Regularly communicate the importance of cybersecurity to the organization's mission, ensuring employees understand how their actions impact overall security.

3. Conduct Continuous Cybersecurity Training and Education

- **Regular Training Programs**: Provide frequent, engaging training sessions that cover current cyber threats, best practices, and the latest security measures. Tailor these sessions to different roles and departments to address specific risks.

- **Hands-on Exercises**: Use simulations, such as phishing drills and incident response scenarios, to help employees recognize and respond effectively to real-world cyber threats.

- **Gamification and Incentives**: Make cybersecurity training interactive and rewarding by incorporating gamification or incentive programs that recognize employees for strong security practices.

4. Foster Open Communication and Reporting

- **Encourage Reporting**: Create an environment where employees feel comfortable reporting potential security incidents, suspicious behavior, or mistakes without fear of blame.

- **Clear Communication Channels**: Provide accessible, confidential channels (e.g., secure email, hotlines) for reporting security concerns and sharing cybersecurity updates.

5. Emphasize Personal Responsibility and Accountability

- **Cybersecurity as Everyone's Responsibility**: Stress that all employees, regardless of their role, contribute to the organization's cybersecurity resilience.

- **Set Expectations for Secure Behavior**: Implement clear policies outlining expected secure behaviors, including secure password management, responsible device usage, and safe data handling.

- **Reinforce Accountability**: Ensure employees understand the consequences of non-compliance with security practices, making accountability part of performance evaluations.

6. Develop and Test Incident Response Plans

- **Regular Incident Response Drills**: Conduct practice exercises involving different departments to prepare for potential security incidents, ensuring everyone knows their roles and responsibilities.

- **Post-Incident Reviews**: After a cyber incident, conduct a detailed review to identify strengths and areas for improvement. Use these findings to refine response protocols and employee training.

7. Implement Role-Based Access and Privilege Management

- **Role-Based Access Control (RBAC)**: Limit system access based on roles, ensuring employees can access only the information necessary for their job functions.

- **Review and Audit Access Regularly**: Periodically audit access controls and privileges to prevent unnecessary or outdated access, reducing the risk of internal threats.

8. Recognize and Reward Good Cybersecurity Practices

- **Incentive Programs**: Recognize employees who demonstrate strong cybersecurity habits, such as reporting threats, participating in training, or following security protocols diligently.

- **Acknowledgment of Cyber Awareness**: Publicly acknowledge cybersecurity achievements or good practices in meetings, newsletters, or reward programs to encourage a positive approach to cybersecurity.

9. Embed Cybersecurity into the Onboarding Process

- **Early Cybersecurity Orientation**: Integrate cybersecurity training into the onboarding process for new employees, setting expectations for security-conscious behavior from day one.

- **Immediate Policy Familiarization**: Familiarize new hires with cybersecurity policies, procedures, and resources so they understand the organization's security standards and their role in maintaining them.

10. Regularly Update Policies and Procedures

- **Review and Adapt Policies**: Cyber threats evolve constantly, so policies and procedures should be regularly reviewed, updated, and communicated to reflect emerging threats and technological advancements.

- **Simplify Policies**: Make cybersecurity policies concise and accessible, ensuring they're easy to understand and apply. Complex policies can discourage adherence.

11. Encourage Collaboration Across Departments

- **Cross-Departmental Collaboration**: Involve various departments in cybersecurity planning and training, from IT and HR to operations and customer service. This helps build a more comprehensive security strategy and promotes a unified approach.

- **Regular Cybersecurity Meetings**: Hold regular interdepartmental meetings to discuss cybersecurity issues, share best practices, and foster a collaborative security culture.

12. Monitor and Assess Cybersecurity Culture Progress

- **Track Culture Metrics**: Use metrics such as incident reports, training participation rates, and compliance audits to assess the effectiveness of cybersecurity culture initiatives.

- **Employee Feedback**: Gather feedback from employees on the cybersecurity training, policies, and practices to identify areas of improvement and make adjustments as needed.

Chapter 12

Incident Response and Crisis Management

Developing An Incident Response Plan for Aviation

Developing an incident response plan for the aviation industry is essential to protect against cyber threats that could compromise passenger safety, operational continuity, and regulatory compliance. A well-crafted incident response (IR) plan provides a structured approach to identify, respond to, and recover from cybersecurity incidents effectively. Here's a step-by-step guide on creating an incident response plan tailored to the unique needs of aviation:

1. Assemble an Incident Response Team (IRT)

- **Define Team Roles**: The team should include cybersecurity experts, IT specialists, legal advisors, public relations personnel, and senior management representatives.

- **Role Specialization**: Assign specific roles, such as incident commander, communications lead, forensic analyst, and technical responders, to handle specialized tasks during an incident.

- **Training and Drills**: Regularly train the team in aviation-specific cyber threats and conduct incident response exercises to ensure preparedness.

2. Identify and Classify Potential Incidents

- **Define Incident Types**: Based on aviation industry risks, define incidents such as:

    o   Unauthorized access to operational systems (avionics, communication, navigation).

- o Breach of passenger or crew data.

- o Airport operational disruptions (e.g., baggage handling, check-in, air traffic management systems).

- **Incident Severity Levels**: Establish severity levels (low, medium, high, critical) to assess an incident's impact on safety, operations, and compliance, ensuring an appropriate response.

3. Detection and Early Identification

- **Set Up Monitoring Tools**: Implement continuous monitoring solutions for aircraft systems, airport networks, and data management systems to detect anomalies.

- **Threat Intelligence and Indicators of Compromise (IOCs)**: Use threat intelligence services and define specific IOCs (e.g., unusual data flows, unauthorized logins) to quickly recognize potential threats.

- **Encourage Reporting**: Create channels for employees, passengers, and partners to report suspicious activity, fostering a proactive security culture.

4. Establish Incident Response Procedures

- **Detailed Response Steps**: For each type of incident, outline a clear series of steps to contain, investigate, and mitigate the impact. Steps might include:

- o Isolating affected systems (aircraft, ground systems, or airport networks).

- o Blocking suspicious IP addresses and disabling compromised accounts.

- o Conducting quick forensic analysis to understand the attack vector.

- **Coordinate with Key Stakeholders**: Ensure that airport authorities, air traffic control (ATC), regulatory bodies, and law enforcement are notified according to the incident's severity.

5. Containment Strategies

- **Short-Term Containment**: Limit the spread of the incident by isolating affected systems or segments of the network. In an aviation context, this may include restricting access to specific onboard or ground systems.

- **Long-Term Containment**: Plan for ongoing monitoring, deploying patches, or updating configurations while keeping essential systems functional and secure.

6. Eradication of Threat

- **Remove Malicious Elements**: Identify and remove malware, backdoors, or unauthorized users from affected systems.

- **System Restoration**: Replace compromised data or systems with backups and verify data integrity.

- **Root Cause Analysis**: Conduct in-depth analysis to understand how the incident occurred and address vulnerabilities to prevent recurrence.

7. Recovery and System Restoration

- **Gradual Restoration of Services**: Ensure affected systems are gradually returned to normal operations. In aviation, prioritize critical systems, like navigation, communication, and airport operations.

- **Testing and Verification**: Conduct thorough testing to confirm systems are fully operational and secure before resuming normal service.

- **Monitor for Recurrence**: Closely monitor systems post-recovery to detect any signs of threat persistence or secondary attacks.

8. Communication and Reporting

- **Internal Communication**: Keep key stakeholders (employees, management, IRT, and potentially partners) informed of the incident status and next steps.

- **Regulatory Notifications**: Notify relevant aviation authorities, such as ICAO, FAA, or EASA, based on incident type and jurisdictional requirements.

- **Public Relations and Passenger Communication**: Have a clear communication plan to address passenger concerns and media inquiries without compromising sensitive information. Be transparent about steps taken to ensure future safety.

9. Documentation and Incident Analysis

- **Detailed Documentation**: Record every action taken during the incident response, including timestamps, communication logs, containment steps, and technical findings.

- **Post-Incident Review**: After containment and recovery, conduct a thorough review with the IR team to identify lessons learned, analyze root causes, and determine areas for improvement.

- **Forensics**: Preserve evidence in cases of potential legal implications, ensuring data integrity for any investigations or legal proceedings.

10. Continuous Improvement and Training

- **Update IR Plan Based on Findings**: Use lessons from each incident to improve procedures, response times, and threat detection capabilities.

- **Regular Drills and Simulations**: Conduct simulated cyber-attacks on various systems (e.g., avionics, ATC) to test and refine the IR plan's effectiveness, preparing the team for real-world incidents.

- **Employee Training and Awareness**: Extend training beyond the IR team to all personnel, especially those who interact with critical systems, teaching them to recognize and report potential security issues.

11. Compliance and Audit Readiness

- **Align with Regulatory Requirements**: Ensure the IR plan adheres to aviation cybersecurity regulations (e.g., ICAO, FAA, EASA) and data protection laws (e.g., GDPR).

- **Documentation for Audits**: Maintain thorough records of incident response activities and periodic reviews, ensuring the organization is prepared for audits and can demonstrate compliance.

Crisis management and communication during a cyberattack in the aviation industry is crucial to contain the impact, reassure stakeholders, and maintain public trust. Due to the potential safety risks and widespread media attention cyber incidents in aviation can attract, a well-prepared crisis management and communication strategy is essential. Here's a comprehensive approach to handling these challenges:

1. Establish a Crisis Management Team (CMT)

- **Define Core Roles**: The crisis management team should include senior executives, cybersecurity leaders, PR and communications experts, legal advisors, and safety officers. In some cases, representatives from regulatory agencies may also be included.

- **Crisis Coordinator**: Assign a designated crisis coordinator responsible for managing the response flow and overseeing communications, ensuring decisions are consistent and timely.

- **Regular Training and Drills**: The team should engage in regular training and simulated crisis drills specific to cyberattacks, focusing on aviation-related scenarios like attacks on air traffic management or passenger data breaches.

2. Develop an Initial Response Protocol

- **Quick Assessment**: Immediately evaluate the scale and impact of the cyberattack to determine if flight safety, passenger data, or operations are compromised.

- **Containment and Control**: Coordinate with the incident response team to isolate affected systems quickly. For example, ground impacted systems, disable specific network connections, or limit access as necessary.

- **Regulatory and Partner Notifications**: Notify key regulatory bodies (e.g., ICAO, FAA, EASA) and industry partners as early as possible if the incident impacts shared systems or public safety.

3. Craft Clear and Consistent Messages

- **Prepare Statements by Impact Area**: Create specific messages for affected areas, such as flight safety, personal data security, or service disruption.

- **Acknowledge and Address the Issue**: Be transparent with stakeholders about the incident's nature, without disclosing sensitive details that could exacerbate the problem or hinder investigations.

- **Emphasize Safety and Security Priorities**: Reassure stakeholders that passenger safety and data security are top priorities, and outline the steps being taken to protect them.

4. Engage in Stakeholder-Specific Communication

- **Internal Communication**: Keep all employees, especially those interacting with passengers and the public, updated on the situation. Empower them with talking points and directions on handling inquiries.

- **Passengers**: If flights or passenger data are impacted, use multiple communication channels (emails, app notifications, airport announcements) to keep passengers informed about disruptions, estimated recovery times, and support options.

- **Regulators and Industry Partners**: Notify regulatory authorities promptly to maintain compliance and provide details about the response and recovery efforts. Industry partners may also need to be informed if shared systems or data are compromised.

- **Media and Public**: Work with the media to share regular updates, addressing the public's concerns while demonstrating the organization's commitment to transparency and swift response.

5. Utilize Multiple Communication Channels

- **Online and Social Media**: Use the company's social media channels, website, and mobile apps to provide real-time updates to the public, keeping messages consistent across platforms.

- **Press Releases and Media Briefings**: Issue press releases and hold regular briefings to provide journalists with accurate, up-to-date information, helping control the narrative and dispel rumors.

- **Crisis Hotline**: Set up a dedicated hotline or online chat support for customers and employees seeking information or assistance.

6. Implement Pre-Crisis Messaging Templates

- **Prepare Template Messages**: Develop message templates for different types of incidents (e.g., data breach, service outage, safety-related attacks) that can be quickly adapted. This saves critical time and helps maintain message clarity.

- **Risk Acknowledgment Statements**: Prepare statements acknowledging known cyber risks, affirming ongoing efforts to enhance cybersecurity, which can be updated as part of real-time communication during the crisis.

7. Ensure Transparency and Timeliness in Updates

- **Set Update Intervals**: Establish consistent intervals for updates (e.g., hourly for high-severity incidents) to ensure stakeholders receive timely information and reduce misinformation risk.

- **Proactive Issue Disclosure**: If an attack involves sensitive data, be proactive about disclosing the nature of compromised information and steps for individuals to protect themselves (e.g., password updates or identity theft monitoring).

8. Manage Rumors and False Information

- **Monitor Social Media and News**: Track social media and news platforms for inaccurate information, rumors, or misleading statements that could damage reputation or escalate public fear.

- **Address Misinformation Swiftly**: Correct any incorrect information quickly and publicly to maintain trust and prevent escalation of concerns.

9. Coordinate with Law Enforcement and Cybersecurity Authorities

- **Engage Authorities Early**: Work with cybersecurity agencies or law enforcement to assist with forensic investigations and, if necessary, pursue criminal charges.

- **Compliance with Investigative Processes**: Ensure that all response actions comply with investigative requirements, including evidence preservation and timely disclosure of relevant findings.

10. Document All Crisis Management Activities

- **Track Key Actions and Decisions**: Maintain detailed records of actions taken, communication issued, and response timelines. This is essential for post-crisis analysis and compliance audits.

- **Incident Log**: Keep a log of key updates, stakeholder communications, and significant findings, which can be useful for both internal evaluation and legal documentation.

11. Post-Crisis Evaluation and Lessons Learned

- **Conduct a Post-Crisis Review**: Once the immediate crisis is resolved, gather the crisis management team to review actions, identify what went well, and discuss areas for improvement.

- **Identify Gaps and Revise Protocols**: Use lessons from the crisis to improve the incident response plan, update communication templates, and strengthen employee training.

- **Report to Stakeholders**: Provide a follow-up report to key stakeholders, including regulatory bodies and customers, outlining the incident's resolution, steps taken, and measures in place to prevent recurrence.

12. Strengthen Employee Training and Awareness

- **Reinforce Cybersecurity Awareness**: Use the incident as an opportunity to raise awareness and train employees on handling cyber threats, reporting suspicious activities, and supporting crisis communication efforts.

- **Incorporate Feedback**: Gather feedback from employees involved in the response to understand challenges faced during the crisis and improve training for future preparedness.

The aviation industry has faced several high-profile cyber incidents where robust incident response played a crucial role in containing threats and restoring normal operations. Here, we examine key case studies that highlight how aviation organizations effectively managed cyber incidents, demonstrating best practices in incident response, containment, communication, and recovery.

1. British Airways Data Breach (2018)

Incident Overview

In 2018, British Airways experienced a data breach that affected approximately 380,000 transactions, compromising customer names, email addresses, and credit card details. The breach was attributed to a sophisticated malware attack on British Airways' website and mobile app.

Response Actions

- **Immediate Detection and Containment**: British Airways quickly identified the breach within a few days of the incident, containing the affected systems to prevent further data exposure.

- **Public Disclosure and Notification**: The company disclosed the breach promptly, informing customers about the compromised information and advising them on protective measures.

- **Collaboration with Authorities**: British Airways worked closely with cybersecurity experts and regulatory authorities to investigate the breach, ensuring a thorough understanding of the attack vector.

- **Enhanced Security Measures**: Following the incident, British Airways implemented stronger security protocols, including multi-layered defenses and continuous monitoring to prevent similar breaches.

Outcome

British Airways' timely incident response minimized further data loss and helped the company take responsibility and reassure customers. Their cooperation with regulatory bodies also reinforced trust and showcased their commitment to data security.

2. Cathay Pacific Data Breach (2018)

Incident Overview

Cathay Pacific discovered unauthorized access to its network, which exposed the personal data of 9.4 million passengers. Information compromised included names, nationalities, dates of birth, passport numbers, and contact details. The breach spanned several months before detection.

Response Actions

- **Swift Containment and Forensics**: Upon detecting unusual activity, Cathay Pacific contained the breach and conducted a forensic investigation to identify the source and scope of the attack.

- **Transparent Public Disclosure**: Although the investigation took time, Cathay Pacific communicated transparently with stakeholders, disclosing details about the breach and the types of data involved.

- **Implementation of Additional Security Measures**: Cathay Pacific upgraded its cybersecurity infrastructure post-incident, introducing new network defenses, access controls, and enhanced monitoring capabilities.

- **Strengthening Internal Processes**: They improved their internal cybersecurity policies and trained staff on heightened security awareness and incident reporting.

Outcome

While the breach affected many passengers, Cathay Pacific's transparent response and subsequent strengthening of security protocols demonstrated accountability and focus on improving security. The incident underscored the importance of swift detection and transparent communication in restoring stakeholder confidence.

3. LOT Polish Airlines DDoS Attack (2015)

Incident Overview

In 2015, LOT Polish Airlines was targeted in a distributed denial-of-service (DDoS) attack that disabled their flight planning system at Warsaw Chopin Airport, grounding approximately 10 flights and delaying dozens more. This incident directly affected flight operations and caused significant disruptions.

Response Actions

- **Rapid Incident Identification and Containment**: LOT's IT team quickly identified the DDoS attack, isolating the affected systems to contain the impact on operations.

- **Immediate Crisis Communication**: The airline communicated proactively with passengers and the public, explaining the cause of the disruptions and setting realistic expectations for resuming operations.

- **Collaboration with Law Enforcement**: LOT coordinated with local authorities and cybersecurity agencies to investigate the source of the DDoS attack and assess further risks.

- **System Hardening**: Following the attack, LOT Polish Airlines strengthened its IT infrastructure, implementing DDoS protection mechanisms to defend against similar attacks in the future.

Outcome

LOT's quick response and transparent communication minimized confusion and maintained passenger trust. Their post-incident infrastructure upgrades also showcased a commitment to future resilience, providing a model for DDoS response in aviation.

4. Air New Zealand Phishing Attack Response (2018)

Incident Overview

Air New Zealand encountered a phishing attack in 2018, in which attackers attempted to compromise employee credentials through deceptive emails. The goal was to gain unauthorized access to sensitive company systems, potentially impacting passenger information and internal operations.

Response Actions

- **Employee Awareness and Training**: Air New Zealand launched a cybersecurity awareness campaign to educate employees on identifying phishing attempts, reporting suspicious emails, and following secure communication practices.

- **Phishing Simulation and Testing**: The airline conducted phishing simulations to test employee readiness and further reinforce security practices.

- **Advanced Monitoring and Detection**: Air New Zealand upgraded its monitoring systems to detect unauthorized access attempts and quickly respond to any anomalies.

- **Policy Adjustments**: Following the incident, Air New Zealand revised its cybersecurity policies to improve response protocols and refine access controls across its systems.

Outcome

Air New Zealand's proactive response included not only resolving the phishing incident but also strengthening its organizational defenses through training and policy improvements. This case highlights the importance of employee awareness in preventing and mitigating phishing-related cyber threats.

5. SITA Data Breach Affecting Multiple Airlines (2021)

Incident Overview

In 2021, SITA, a global IT provider to the air transport industry, experienced a data breach impacting passenger service systems. Airlines including Singapore Airlines, Lufthansa, and Finnair were affected, with the breach compromising data of an estimated 4.5 million passengers.

Response Actions

- **Cross-Industry Coordination**: SITA coordinated closely with its client airlines to assess the impact and inform affected passengers. Each airline independently communicated the breach to their passengers, fostering transparency and consistency.

- **Prompt Notification**: Both SITA and the affected airlines promptly notified the public and regulators, complying with data protection requirements.

- **Industry-Wide Risk Assessment and Security Enhancements**: Following the breach, SITA and several airlines undertook a comprehensive review of third-party vendor security practices to strengthen supply chain cybersecurity.

- **Improved Vendor Management**: The incident emphasized the need for regular audits and compliance reviews for vendors, ensuring their security standards align with the aviation sector's requirements.

Outcome

The collaborative response across SITA and affected airlines demonstrated effective crisis management within a multi-stakeholder environment. Their transparency, combined with industry-wide security improvements, helped maintain public trust in aviation cybersecurity.

Chapter 13

Future of Cybersecurity in Aviation

Emerging Threats and New Types of Cyberattacks

As technology in the aviation industry becomes more sophisticated and interconnected, new types of cyber threats are emerging, posing significant risks to the safety, security, and efficiency of airline operations, passenger data, and infrastructure. Here, we explore some of the evolving cyber threats that are reshaping the aviation cybersecurity landscape and discuss the unique challenges they present.

1. Artificial Intelligence-Driven Attacks

Description: Cybercriminals are increasingly using artificial intelligence (AI) to create more effective and adaptive attacks. AI can enable attackers to craft highly targeted phishing schemes, launch sophisticated malware, and rapidly exploit vulnerabilities in aviation systems. AI-based attacks can learn from previous security protocols, adapt to new defenses, and evade traditional cybersecurity measures.

Challenges:

- AI-driven malware can evolve autonomously, making it difficult to detect with conventional security tools.

- AI-powered phishing can bypass human detection by mimicking legitimate messages with a high degree of accuracy.

- This type of attack can create complex threats that are challenging to counter in real-time, especially across highly interconnected aviation networks.

2. Ransomware Targeting Critical Systems

Description: Ransomware attacks are growing in both frequency and sophistication. These attacks encrypt critical systems or data, halting operations until a ransom is paid. In the aviation industry, ransomware could target airline reservation systems, air traffic control networks, or critical airport infrastructure, leading to massive disruptions.

Challenges:

- Disruption to real-time operations (flight schedules, baggage handling, etc.) can lead to severe financial and reputational damage.

- The industry's reliance on continuous operations makes it more susceptible to ransom payments to quickly restore normalcy.

- Ransomware attacks could lead to prolonged downtime in essential aviation services, affecting the entire travel ecosystem.

3. Attacks on Internet of Things (IoT) Devices

Description: The rise of IoT in aviation—such as smart baggage tracking, connected cockpit systems, and maintenance sensors—introduces many new entry points for cyber attackers. These devices, often with minimal security protocols, can be hijacked or tampered with, potentially compromising the safety and security of flights and airport operations.

Challenges:

- IoT devices are often not designed with strong cybersecurity measures, making them easy targets for attackers.

- Many IoT devices transmit sensitive data across networks, creating opportunities for man-in-the-middle attacks and data interception.

- Compromised IoT devices could be used as an entry point to infiltrate larger networks, impacting everything from aircraft systems to airport facilities.

4. Supply Chain Attacks

Description: Attackers increasingly target third-party vendors and service providers to gain access to aviation networks. This can include software providers, maintenance companies, and technology partners. By compromising suppliers, attackers can bypass traditional security protocols and directly impact the operations of airlines and airports.

Challenges:

- The interconnected nature of aviation operations means that vulnerabilities in one partner can have widespread impacts.

- Monitoring the cybersecurity practices of all third-party providers is challenging and requires constant vigilance.

- Supply chain attacks can be difficult to detect as they often exploit trust relationships within aviation's complex ecosystem.

5. Cloud Vulnerabilities

Description: With many aviation organizations moving to cloud-based infrastructure, cloud vulnerabilities are emerging as a significant threat. Data breaches, misconfigured cloud services, and insecure access protocols in the cloud environment can expose sensitive passenger information, operational data, and financial records.

Challenges:

- Shared cloud infrastructure makes it easier for attackers to breach multiple systems if they can compromise one layer.

- Misconfigured cloud services are a common vulnerability, often due to the complexity of managing cloud security settings.

- Cloud-based attacks can lead to data exfiltration or deletion, severely affecting airlines' ability to operate.

6. Quantum Computing Threats to Encryption

Description: Although still in its infancy, quantum computing poses a future threat to encryption. Quantum computers could potentially break current encryption algorithms quickly, making it easier for attackers to access sensitive data, including passenger records and operational commands.

Challenges:

- Transitioning to quantum-resistant encryption is complex and requires large-scale changes across the aviation sector.

- Highly sensitive data like crew credentials and passenger information is vulnerable if encryption standards become obsolete.

- Developing quantum-resistant algorithms and protocols is essential to mitigate future risks from quantum computing.

7. Deepfake-Based Social Engineering

Description: Deepfake technology can be used to create realistic audio and video content, enabling attackers to impersonate executives or employees within aviation organizations. This can facilitate phishing attempts or fraudulent instructions, leading to unauthorized access to systems or sensitive information.

Challenges:

- Deepfake-based social engineering attacks are difficult to detect, as they appear highly authentic.

- They can be used to manipulate staff into revealing sensitive information or performing unauthorized actions.

- Aviation companies must invest in training and verification processes to recognize and counter these sophisticated scams.

8. Cyber-Physical Attacks on Operational Technology (OT)

Description: Operational technology (OT) controls physical processes, such as baggage systems, air conditioning, and fuel management. Cyber-physical attacks target these systems to disrupt operations or even endanger passenger safety.

Challenges:

- OT systems often lack the same level of cybersecurity as IT systems, making them easier targets.

- A compromised OT system can have direct, tangible impacts on passengers and operations, posing serious safety risks.

- The integration of OT and IT systems in airports and aircraft introduces vulnerabilities where an IT attack can affect physical operations.

9. Insider Threats Enhanced by Remote Work

Description: As remote work becomes more prevalent, aviation employees working offsite may lack secure access protocols, which increases the risk of insider threats. Malicious insiders or unwitting employees with compromised devices can introduce vulnerabilities into aviation networks.

Challenges:

- Remote access introduces more potential entry points for attackers, requiring enhanced security monitoring.

- Remote work policies must include stringent cybersecurity measures to prevent unauthorized access.

- Insider threats are difficult to detect and require comprehensive monitoring, especially when employees are working outside secured environments.

10. Disruption of Satellite and GPS Systems

Description: Many aviation systems rely on satellite-based navigation and GPS technology for flight operations. Jamming or spoofing these signals could cause flight route disruptions, misdirect aircraft, or interfere with air traffic control communications.

Challenges:

- Interference with GPS or satellite signals can have immediate, operational impacts on flight safety and routing.

- Protecting these systems requires specialized countermeasures and collaboration with global partners.

- Detection of spoofing or jamming is challenging, requiring advanced tools and continuous monitoring to ensure accuracy.

As aviation technology continues to advance, the industry's dependence on digital systems creates new and complex cybersecurity challenges. Predicting and preparing for these future challenges is critical to safeguarding the aviation sector. This section highlights key cybersecurity challenges anticipated in the coming years and outlines the implications for airlines, airports, air traffic management, and passenger safety.

1. Increasing Complexity of Cyber-Physical Integration

Description: Modern aircraft and airport operations rely heavily on cyber-physical systems, where IT (Information Technology) and OT (Operational Technology) are tightly integrated. This integration enables more efficient operations and data-sharing, but it also introduces new vulnerabilities that can be exploited by cyber attackers.

Challenges:

- Managing and securing the interaction between IT and OT systems will require more advanced cybersecurity frameworks.

- Attackers may leverage the interconnected nature of these systems to disrupt physical operations, such as baggage handling, fuel management, or even in-flight controls.

- Ensuring the security of legacy OT systems, which often lack robust cybersecurity protections, will be increasingly challenging.

2. Quantum Computing and Cryptography Threats

Description: Quantum computing has the potential to break current encryption standards, rendering traditional cryptographic protections obsolete. Although quantum computers are not yet widely available, rapid advancements could make them accessible within a decade, posing a significant risk to aviation cybersecurity.

Challenges:

- Transitioning to quantum-resistant encryption will be complex and costly for airlines, airports, and associated systems.

- Ensuring data integrity, especially in real-time communications like air-to-ground and air traffic control, will require substantial changes.

- Developing and implementing quantum-safe cryptography across all aviation systems will require industry-wide collaboration.

3. Expanding Attack Surface with Internet of Things (IoT) and 5G

Description: The use of IoT devices in aviation is expanding, from connected cockpit systems to airport infrastructure and passenger applications. The rollout of 5G networks will enable faster, more reliable connections, further increasing the number of connected devices.

Challenges:

- As more IoT devices are integrated, each becomes a potential entry point for cyber threats.

- Many IoT devices lack robust security, and securing such a vast array of devices will require unified security standards and practices.

- 5G technology introduces new networking vulnerabilities and the possibility of exploitation through sophisticated attacks that target the high-speed network itself.

4. Sophisticated AI and Machine Learning Attacks

Description: While AI and machine learning (ML) have been instrumental in enhancing cybersecurity, they can also be used by attackers to create more adaptive and autonomous threats. AI-driven attacks could include highly targeted phishing, advanced malware, or autonomous botnet attacks designed to evade traditional defenses.

Challenges:

- Detecting AI-driven attacks will require advanced AI-based defense mechanisms capable of identifying subtle patterns in real time.

- Adaptive malware powered by machine learning could circumvent defenses by continuously changing its behavior.

- The potential of AI to launch autonomous cyberattacks requires cybersecurity systems that can respond at machine speed.

5. Emergence of Smart Airports and Infrastructure Vulnerabilities

Description: As airports adopt more "smart" technologies—like automated passenger check-ins, biometric identification, and AI-enhanced surveillance—they become more efficient but also more vulnerable. Smart airports rely on interconnected digital systems that are targets for cyber attackers looking to disrupt operations or steal sensitive data.

Challenges:

- Securing interconnected systems, especially when integrating third-party technology, will require unified standards and rigorous security protocols.

- Cyber threats to smart infrastructure can disrupt not just IT systems but physical operations, causing significant operational and safety risks.

- Balancing efficiency with security in smart airports will be challenging, as faster processes must be secured without compromising passenger experience.

6. Increasing Risk of Insider Threats and Social Engineering

Description: Insider threats remain a critical concern, especially as the industry shifts to remote and hybrid work environments. Social engineering attacks that manipulate employees into providing sensitive information or access are also on the rise, and as cyber tactics become more sophisticated, insiders may be increasingly exploited.

Challenges:

- Insider threats are often difficult to detect, as malicious activities may look like normal operations.

- Strengthening employee training against social engineering attacks, particularly phishing, will be essential.

- Remote work introduces new risks for aviation personnel who may access sensitive data from less secure environments, requiring advanced monitoring and security protocols.

7. Ransomware Targeting Critical Aviation Infrastructure

Description: Ransomware attacks have escalated across industries, with cybercriminals targeting critical infrastructure to maximize impact and compel ransom payments. In aviation, ransomware could disrupt airport operations, air traffic control, or airline reservation systems, leading to costly delays and security risks.

Challenges:

- The downtime caused by ransomware attacks can be especially detrimental to the aviation industry, which relies on continuous operations.

- Backup and recovery plans must be optimized, as recovering systems from ransomware attacks can be time-consuming and costly.

- Preventing ransomware requires proactive cybersecurity measures and regular training of personnel to prevent phishing, the most common method of entry.

8. Regulatory Pressure and Compliance with Global Standards

Description: As cybersecurity threats escalate, regulators are implementing stricter standards for aviation cybersecurity. International bodies such as ICAO, FAA, and EASA are developing cybersecurity regulations, requiring airlines, airports, and air traffic management to implement minimum cybersecurity measures and report incidents.

Challenges:

- Compliance with multiple national and international standards can be costly and complex for global aviation organizations.

- Regulatory requirements often lag behind emerging threats, creating gaps in security.

- Ensuring that cybersecurity measures are consistent across global operations will require comprehensive training and implementation strategies.

9. Supply Chain Vulnerabilities and Third-Party Risks

Description: Aviation relies on an extensive network of third-party vendors for everything from software and hardware to maintenance and logistics. A cyber incident affecting one vendor can disrupt entire operations, making supply chain cybersecurity a top concern.

Challenges:

- Monitoring the cybersecurity practices of every vendor in the supply chain is challenging and resource-intensive.

- Many aviation organizations lack transparency into the cybersecurity posture of third-party suppliers, increasing the risk of unanticipated vulnerabilities.

- Strengthening supply chain security will require robust screening processes and ongoing vendor management practices.

10. Resilience Against Large-Scale Distributed Denial of Service (DDoS) Attacks

Description: DDoS attacks are increasing in scale and sophistication, targeting aviation networks, airline websites, and air traffic management systems. These attacks can disrupt services, causing delays and affecting customer experience, especially during peak travel times.

Challenges:

- DDoS attacks can cause widespread disruption by overloading network infrastructure, leading to operational slowdowns.

- Defending against large-scale DDoS attacks requires both technological investments in mitigation tools and coordinated response planning.

- Preparing for DDoS attacks includes resilience planning to ensure continuity of service even under extreme network stress.

In the face of rapidly advancing cyber threats, the aviation industry must embrace innovation and develop proactive strategies to anticipate and mitigate risks before they can cause significant disruption. From adopting cutting-edge technology to fostering industry-wide collaboration, this section explores various approaches that can help aviation organizations stay ahead of cyber threats and enhance the resilience of their systems.

1. Proactive Threat Intelligence and Monitoring

Overview: Proactive threat intelligence involves gathering and analyzing data on potential threats and attack patterns before they can impact aviation systems. By adopting a predictive, intelligence-driven approach, airlines, airports, and air traffic management systems can identify emerging risks and vulnerabilities in real time.

Strategies:

- Invest in AI-driven threat intelligence platforms that use machine learning to analyze patterns and detect anomalies.

- Participate in threat intelligence sharing networks like the Aviation ISAC (Information Sharing and Analysis Center) to gain insights on emerging threats across the aviation industry.

- Develop a 24/7 Security Operations Center (SOC) to monitor systems continuously and respond to incidents before they escalate.

2. Advancing Cybersecurity with Artificial Intelligence (AI) and Machine Learning (ML)

Overview: AI and ML technologies play a vital role in identifying and countering sophisticated cyber threats. By processing large datasets quickly, these tools can detect unusual behaviors and anomalies that may indicate an impending attack, helping security teams respond faster.

Strategies:

- Implement machine learning algorithms for real-time anomaly detection in networks, allowing rapid identification of potential threats.

- Utilize AI-driven behavioral analytics to detect unusual access patterns, especially in critical systems like air traffic management and in-flight networks.

- Invest in ML-powered automated response tools that can mitigate low-level threats without human intervention, freeing up resources for more complex issues.

3. Adopting Quantum-Safe Encryption

Overview: As quantum computing technology progresses, traditional encryption methods may become vulnerable to decryption by quantum computers. Transitioning to quantum-safe encryption protocols will be essential to protect sensitive data in the future.

Strategies:

- Begin adopting quantum-resistant cryptographic algorithms to secure data in communication systems, from passenger information to air traffic control exchanges.

- Collaborate with encryption specialists and quantum computing experts to prepare and test encryption protocols that can withstand quantum attacks.

- Conduct a phased rollout of quantum-safe encryption across all aviation systems, prioritizing the most critical data and communication channels.

4. Developing Cyber-Resilient Infrastructure and Systems

Overview: Cyber-resilience refers to an organization's ability to maintain critical operations despite cyberattacks. Building resilient infrastructure helps aviation systems recover quickly from disruptions, minimizing the impact on operations and passengers.

Strategies:

- Design redundancy into critical systems like air traffic control, passenger management, and baggage handling, ensuring backup capabilities in case of system compromise.

- Regularly test disaster recovery plans and incident response protocols through simulations and drills to identify and address any vulnerabilities.

- Implement compartmentalized architecture within networks, limiting the ability of attackers to move laterally and minimizing the impact of breaches on interconnected systems.

5. Embracing Zero Trust Architecture

Overview: Zero Trust is a security framework that assumes no user or device should be trusted by default, even if they are within the network. This approach is especially useful in aviation, where numerous interconnected systems and devices require strict access control.

Strategies:

- Establish identity and access management (IAM) systems that enforce strict authentication and authorization protocols for all users and devices.

- Implement network segmentation and micro-segmentation within airport and airline systems, creating barriers that prevent intruders from accessing critical assets.

- Deploy continuous monitoring and validation tools that verify user identities and device compliance with security policies at every point of access.

6. Strengthening Supply Chain Security

Overview: Aviation relies on a vast network of third-party vendors, from software providers to equipment manufacturers. Securing the supply chain ensures that these partners do not introduce vulnerabilities into aviation systems.

Strategies:

- Conduct thorough cybersecurity audits of suppliers and require them to meet industry standards, such as ISO 27001 for information security management.

- Establish contractual requirements for third-party vendors to report cyber incidents, undergo regular security assessments, and implement strong security practices.

- Invest in continuous monitoring of supply chain cybersecurity, identifying any risks that could affect operational safety or data security.

7. Enhancing Cybersecurity Training and Awareness

Overview: Human error remains a significant cause of cybersecurity incidents. Comprehensive training and awareness programs can reduce this risk by equipping aviation personnel with the knowledge and skills needed to recognize and respond to cyber threats.

Strategies:

- Provide cybersecurity training tailored to different roles within the aviation sector, focusing on the specific risks faced by airport staff, pilots, air traffic controllers, and administrative personnel.

- Implement phishing simulation programs to educate employees on identifying and avoiding phishing scams, which are common in aviation cyberattacks.

- Create a cybersecurity culture where employees feel empowered to report suspicious activities and understand their role in safeguarding the organization.

8. Leveraging Blockchain for Data Integrity and Transparency

Overview: Blockchain technology, with its decentralized and tamper-proof nature, offers unique advantages for data integrity in aviation. It can be used to enhance data security and transparency, especially in areas like passenger information, supply chain records, and maintenance logs.

Strategies:

- Explore blockchain-based identity management systems to securely verify passenger and crew identities, reducing identity fraud risks.

- Implement blockchain solutions for tracking and verifying aircraft maintenance and repair records, enhancing the security and transparency of data shared between airlines, airports, and regulators.

- Use blockchain for secure and transparent tracking of supply chain data, reducing the risk of tampering and fraud among vendors and suppliers.

9. Implementing Behavioral Analytics to Detect Insider Threats

Overview: Insider threats, whether intentional or accidental, pose a unique risk to aviation security. Behavioral analytics tools can help detect unusual or unauthorized actions by employees, contractors, or partners.

Strategies:

- Deploy behavioral analytics software that continuously monitors user activities and flags anomalies in real time.

- Integrate behavior monitoring into access control systems, so any deviations from typical behavior trigger alerts for further investigation.

- Conduct regular reviews and analysis of access logs to identify trends that could indicate potential insider threats.

10. Promoting Collaboration and Information Sharing Across the Industry

Overview: Cybersecurity in aviation benefits from collaborative efforts, as threat intelligence and response strategies are stronger when shared across industry stakeholders, government agencies, and international organizations.

Strategies:

- Join and actively participate in aviation-specific information-sharing forums, like the Aviation ISAC, to stay informed on the latest threats and vulnerabilities.

- Partner with government agencies and cybersecurity research organizations to share best practices, threat intelligence, and response tactics.

- Collaborate with global regulatory bodies (e.g., ICAO, EASA, FAA) to harmonize cybersecurity standards and policies, creating a unified approach to aviation security.

11. Investing in Continuous Innovation and R&D

Overview: The pace of technological evolution requires the aviation industry to stay ahead by continually investing in research and development of new cybersecurity solutions.

Strategies:

- Allocate budget and resources for R&D focused on emerging cybersecurity technologies, such as AI-enhanced threat detection, quantum-safe cryptography, and resilient infrastructure.

- Encourage partnerships with academic institutions, startups, and cybersecurity firms that are developing cutting-edge solutions for aviation.

- Support innovation initiatives, such as hackathons or collaborative security projects, that can explore novel approaches to solving complex cybersecurity challenges in aviation.

Chapter 14

Securing the Skies for Tomorrow

Key Cybersecurity Points and Their Importance

As the aviation industry becomes increasingly digital and interconnected, cybersecurity emerges as a paramount concern. The following key points encapsulate the essential aspects of aviation cybersecurity and their significance in shaping the future of air travel:

1. Comprehensive Understanding of Aircraft Systems and Components

- **Key Points**: Modern aircraft incorporate sophisticated systems like flight controls, navigation, communication, avionics, and environmental controls.

- **Importance**: A deep understanding of these systems is crucial for identifying potential cybersecurity vulnerabilities that could impact flight safety and operational efficiency.

2. Identification of Vulnerabilities in Avionics, Communication, and Navigation Systems

- **Key Points**: Vulnerabilities include software flaws, outdated systems, unencrypted communications, and integration with IoT devices.

- **Importance**: Addressing these vulnerabilities is vital to prevent unauthorized access, data manipulation, and operational disruptions that could jeopardize passenger safety and airline operations.

3. In-Flight Cybersecurity Concerns and Scenarios

- **Key Points**: Risks involve passenger Wi-Fi networks, interconnected systems, avionics manipulation, and remote access threats.

- **Importance**: Ensuring robust in-flight cybersecurity measures protects critical systems from being compromised, maintaining the integrity and safety of flights.

4. Cyber Threats Specific to Airport Operations

- **Key Points**: Threats include data breaches, ransomware, DoS attacks, insider threats, and compromised supply chains.

- **Importance**: Securing airport operations safeguards sensitive data, ensures smooth operations, and maintains passenger trust and safety.

5. Vulnerabilities in Ground Systems: Baggage Handling, Check-In, and Surveillance

- **Key Points**: Automation, legacy systems, physical access, and insecure surveillance systems are primary vulnerabilities.

- **Importance**: Protecting these ground systems prevents operational disruptions, data breaches, and security compromises, enhancing overall airport functionality.

6. Protecting Data and Personal Information in Airports

- **Key Points**: Challenges include data volume, interconnected systems, third-party risks, and insider threats.

- **Importance**: Implementing robust data protection strategies ensures compliance with regulations, maintains customer trust, and prevents identity theft and fraud.

7. Compliance with Data Protection Laws (e.g., GDPR, CCPA)

- **Key Points**: Regulations mandate secure data handling, breach reporting, and protection of personal information.

- **Importance**: Compliance avoids hefty penalties, legal repercussions, and reputational damage, while fostering a culture of data privacy and security.

8. Strategies for Safeguarding Sensitive Information

- **Key Points**: Includes data classification, access control, encryption, regular security assessments, data minimization, and employee training.

- **Importance**: These strategies collectively enhance data security, reduce exposure to threats, and ensure the integrity and confidentiality of sensitive information.

9. Cybersecurity Regulations and Standards in Aviation (ICAO, EASA, FAA)

- **Key Points**: International and national bodies establish frameworks and guidelines to manage cybersecurity risks.

- **Importance**: Adhering to these regulations ensures standardized security practices, facilitates international collaboration, and enhances overall aviation security.

10. Emerging Policies and the Role of International Cooperation

- **Key Points**: Policies focus on national strategies, regulatory frameworks, data protection, and global collaboration.

- **Importance**: International cooperation harmonizes cybersecurity efforts, shares threat intelligence, and fosters a unified defense against global cyber threats.

11. Key Cybersecurity Technologies: Encryption, Firewalls, and Intrusion Detection Systems (IDS)

- **Key Points**: These technologies protect data integrity, control access, and detect malicious activities.

- **Importance**: Implementing advanced cybersecurity technologies forms the backbone of defense mechanisms, safeguarding aviation systems from unauthorized access and cyberattacks.

12. Role of Artificial Intelligence (AI) and Machine Learning (ML) in Threat Detection

- **Key Points**: AI and ML enhance threat detection through anomaly detection, behavioral analytics, and automated responses.

- **Importance**: Leveraging AI and ML improves the efficiency and accuracy of identifying and mitigating cyber threats, enabling proactive security measures.

13. Advances in Quantum Computing and Its Impact on Aviation Security

- **Key Points**: Quantum computing poses future threats to current encryption methods but also offers new security solutions like quantum key distribution.

- **Importance**: Preparing for quantum threats ensures long-term data security, while adopting quantum-safe technologies strengthens defense against next-generation cyberattacks.

14. Cyber Hygiene Practices for Airlines, Airports, and Passengers

- **Key Points**: Involves regular system updates, strong authentication, secure data handling, and passenger awareness.

- **Importance**: Good cyber hygiene minimizes vulnerabilities, prevents breaches, and promotes a secure environment for all stakeholders in aviation.

15. Importance of Cybersecurity Training for Aviation Personnel

- **Key Points**: Training enhances awareness, reduces human error, ensures regulatory compliance, and strengthens incident response capabilities.

- **Importance**: A well-trained workforce is the first line of defense against cyber threats, fostering a security-conscious culture and improving overall organizational resilience.

16. Strategies for Building a Resilient Cybersecurity Culture

- **Key Points**: Involves leadership commitment, continuous training, open communication, accountability, and fostering collaboration.

- **Importance**: A resilient cybersecurity culture ensures that security practices are ingrained in daily operations, enabling organizations to swiftly respond to and recover from cyber incidents.

17. Developing an Incident Response Plan for Aviation

- **Key Points**: Includes assembling an incident response team, defining procedures, establishing communication protocols, containment strategies, eradication, recovery, and post-incident analysis.

- **Importance**: A robust incident response plan minimizes the impact of cyberattacks, ensures swift recovery, and maintains operational continuity and safety.

18. Crisis Management and Communication During a Cyberattack

- **Key Points**: Effective crisis management involves clear communication, stakeholder engagement, collaboration with authorities, and transparent reporting.

- **Importance**: Managing communication during a cyber crisis preserves trust, reduces panic, and ensures that all parties are informed and coordinated in response efforts.

19. Case Studies of Effective Incident Response in Aviation

- **Key Points**: Examples include British Airways, Cathay Pacific, LOT Polish Airlines, Air New Zealand, and SITA, demonstrating successful containment, communication, and recovery.

- **Importance**: These case studies provide valuable lessons and best practices, showcasing the effectiveness of proactive incident response strategies and highlighting areas for improvement.

20. Emerging Threats and New Types of Cyberattacks

- **Key Points**: Includes AI-driven attacks, ransomware targeting critical systems, IoT vulnerabilities, supply chain attacks, quantum computing threats, deepfake-based social engineering, cyber-physical attacks, insider threats, and DDoS attacks.

- **Importance**: Understanding and preparing for these evolving threats ensures that aviation organizations can adapt their defenses and maintain robust security in a dynamic threat landscape.

21. Predicting Future Challenges in Cybersecurity for Aviation

- **Key Points**: Future challenges include increasing cyber-physical integration, quantum computing, expanding IoT and 5G attack surfaces, sophisticated AI-driven threats, smart airport vulnerabilities, insider threats, ransomware targeting critical infrastructure, regulatory pressures, supply chain vulnerabilities, and DDoS resilience.

- **Importance**: Anticipating these challenges allows aviation stakeholders to proactively develop strategies, invest in necessary technologies, and establish robust security measures to maintain safety and operational integrity.

22. Innovation and Strategies to Stay Ahead of Evolving Threats

- **Key Points**: Embracing AI and ML, adopting quantum-safe encryption, developing cyber-resilient infrastructure, implementing Zero Trust architectures, enhancing supply chain security, leveraging blockchain, fostering collaboration and information sharing, continuous innovation and R&D, and strengthening cybersecurity training.

- **Importance**: Innovative approaches and forward-thinking strategies ensure that the aviation industry remains resilient against emerging cyber threats, maintaining the highest standards of safety, security, and operational efficiency.

Overall Importance for the Future of Aviation

1. **Safety and Security**: Robust cybersecurity measures directly correlate with passenger and crew safety, protecting against threats that could compromise flight operations and airport infrastructure.

2. **Operational Continuity**: Effective cybersecurity ensures that critical systems remain functional, minimizing

disruptions caused by cyberattacks and maintaining seamless aviation operations.

3. **Regulatory Compliance**: Adhering to international and national cybersecurity regulations safeguards aviation organizations from legal penalties and enhances their reputation.

4. **Customer Trust and Reputation**: Protecting personal data and maintaining secure operations builds and sustains passenger trust, which is essential for business success and industry reputation.

5. **Technological Advancement**: Staying ahead of cyber threats through innovation ensures that the aviation industry can leverage new technologies safely, enhancing efficiency and service quality.

6. **Resilience and Adaptability**: A resilient cybersecurity culture enables aviation organizations to adapt to evolving threats, ensuring long-term sustainability and robustness against future challenges.

7. **Global Collaboration**: International cooperation and information sharing foster a unified defense against global cyber threats, enhancing the overall security posture of the aviation sector worldwide.

As the aviation industry rapidly advances in technology, the emphasis on cybersecurity is essential for creating a safer, more resilient aviation environment. Cybersecurity in aviation goes beyond protecting data; it encompasses the safety of flights, the security of airport operations, and the assurance of seamless air traffic management. Here are some key strategies for achieving a safer aviation industry through cybersecurity:

1. Integrating Security into Aircraft Design and Development

- **Strategy**: Cybersecurity must be embedded into the design and development phases of aircraft systems. Avionics, communication, and navigation systems should be built with security as a core feature.

- **Impact**: This approach creates more resilient aircraft systems less vulnerable to cyber threats and able to withstand potential attacks.

2. Strengthening Data Protection for Passengers and Crew

- **Strategy**: Implementing robust data protection measures like encryption, access control, and data anonymization safeguards passenger and crew data throughout its lifecycle.

- **Impact**: Protecting sensitive information maintains customer trust and complies with global regulations (e.g., GDPR, CCPA), fostering a culture of data privacy in the aviation industry.

3. Enhancing Airport Cybersecurity

- **Strategy**: Airports must secure their infrastructure, from surveillance systems to baggage handling and check-in services, by addressing vulnerabilities in operational technology (OT) and information technology (IT).

- **Impact**: By reducing the risk of cyber disruptions, airports can ensure smooth, uninterrupted operations, enhancing overall passenger experience and safety.

4. Securing Air Traffic Management (ATM) Systems

- **Strategy**: Air traffic management systems should use advanced encryption, secure communication protocols, and redundancy to protect air-to-ground and ground-to-ground communications.

- **Impact**: Securing ATM systems ensures accurate communication and coordination between pilots and air traffic controllers, reducing the risk of miscommunications and potential incidents.

5. Addressing In-Flight Cybersecurity Concerns

- **Strategy**: Implementing network segmentation and access control measures within aircraft protects critical avionics from potential threats stemming from passenger Wi-Fi and other in-flight services.

- **Impact**: This segmentation minimizes the risk of cyber threats affecting flight control systems, preserving the integrity and safety of in-flight operations.

6. Building a Cyber-Resilient Workforce

- **Strategy**: Regular cybersecurity training for pilots, ground crew, air traffic controllers, and airport staff ensures that personnel can recognize and respond to cyber threats effectively.

- **Impact**: A well-trained workforce acts as a first line of defense, able to detect and respond to cyber incidents, reducing the likelihood of breaches due to human error.

7. Adopting Key Cybersecurity Technologies

- **Strategy**: Technologies like firewalls, intrusion detection systems (IDS), encryption, and advanced endpoint security tools protect against unauthorized access and data breaches.

- **Impact**: Implementing these technologies strengthens defenses against common cyber threats such as malware, ransomware, and unauthorized data access, ensuring system integrity.

8. Leveraging Artificial Intelligence (AI) and Machine Learning (ML)

- **Strategy**: AI and ML can enhance threat detection and response through real-time monitoring, anomaly detection, and automated incident response.

- **Impact**: With AI and ML, aviation systems can identify unusual activity and potential cyber threats proactively, allowing faster, more accurate responses to cyber incidents.

9. Promoting Cyber Hygiene and Awareness

- **Strategy**: Educating passengers on secure practices, such as protecting their personal devices and being cautious with public Wi-Fi, supports cybersecurity at all levels.

- **Impact**: Improved cyber hygiene reduces the risk of malware spreading from passenger devices to airline networks, further safeguarding the aviation ecosystem.

10. Developing a Comprehensive Incident Response Plan

- **Strategy**: Having a well-defined incident response plan (IRP) with clear roles, protocols, and communication channels enables swift action during a cyber crisis.

- **Impact**: A robust IRP minimizes damage from cyber incidents, reduces recovery time, and maintains operational continuity, protecting passengers, crew, and infrastructure.

11. International Cooperation and Information Sharing

- **Strategy**: Collaborating on cybersecurity standards, sharing threat intelligence, and creating alliances across countries promotes a unified defense strategy.

- **Impact**: International cooperation enables faster response to global threats, ensures consistent cybersecurity practices, and fosters a more secure and connected aviation industry.

12. Continuous Innovation and Investment in Cybersecurity

- **Strategy**: Ongoing research and investment in cybersecurity, including exploring emerging technologies like quantum-safe encryption, ensures that the aviation industry stays ahead of new threats.

- **Impact**: Proactive investment in cybersecurity helps aviation stakeholders remain resilient in a rapidly evolving threat landscape, building a foundation for future-safe air travel.

Long-Term Impact: A Safer, Trustworthy Aviation Industry

Through a commitment to cybersecurity, the aviation industry can reduce cyber risks, prevent costly incidents, and build passenger and stakeholder confidence. A safer aviation industry ultimately benefits everyone, ensuring reliable, secure, and efficient air travel experiences. By adopting these strategies, aviation leaders are not only protecting the skies but also setting a global standard for cybersecurity in critical infrastructure.

The aviation industry is at a pivotal moment where digital transformation meets unprecedented cybersecurity challenges. To protect the skies, we need a collective commitment to strengthen cyber defenses across all sectors of aviation—airlines, airports, manufacturers, and regulatory bodies. Cyber threats evolve rapidly, and so must our strategies for addressing them. This journey demands a united effort toward continuous improvement, proactive investment, and an unwavering dedication to cybersecurity at every level.

1. Build a Culture of Shared Responsibility

Every stakeholder in the aviation ecosystem has a role in enhancing cybersecurity. Airlines, airports, technology providers, regulators, and passengers must work together to protect critical infrastructure. This requires:

- **Collaborative training and awareness** initiatives to instill cybersecurity best practices among aviation personnel, contractors, and the public.

- **Unified policies and procedures** that ensure security standards are consistently applied across borders and organizations.

2. Encourage Cross-Border Collaboration and Information Sharing

Cyber threats in aviation are not confined to national borders. To stay ahead, countries and organizations must prioritize:

- **Information sharing** networks to quickly relay threat intelligence, vulnerabilities, and attack patterns across the globe.

- **International partnerships** that establish joint protocols for incident response and recovery, allowing for swift action on emerging threats.

3. Drive Innovation in Cybersecurity Technology

With rapid advancements in cyber tactics, the aviation sector must continuously innovate to stay ahead. Key areas for investment include:

- **Advanced AI and machine learning** systems for real-time threat detection and automated response.

- **Quantum-resistant encryption** to secure sensitive data and communications as quantum computing capabilities develop.

- **Resilient network architectures** that limit potential impact areas and create redundancies to sustain operations during attacks.

4. Regularly Update and Test Incident Response Plans

Preparedness is a cornerstone of effective cybersecurity. Airlines, airports, and regulators must:

- **Continuously review and refine** incident response plans to address new threats and technological changes.

- **Conduct regular simulations and drills** that involve all relevant personnel, ensuring that teams are prepared to act quickly and efficiently under pressure.

- **Implement post-incident reviews** to learn from each event and improve future responses.

5. Champion Cybersecurity Standards and Regulations

Unified standards and strong regulatory frameworks are essential for maintaining a secure global aviation industry. Stakeholders should:

- **Advocate for and adhere to international standards**, such as those set by ICAO, EASA, and FAA, which help ensure consistent security measures across countries and organizations.

- **Participate in standard-setting bodies and policy forums** to address emerging challenges and incorporate best practices into evolving regulations.

- **Support harmonized cybersecurity guidelines** that align across regions, allowing for coherent security practices that bolster global travel safety.

www.ingramcontent.com/pod-product-compliance
Lightning Source LLC
Chambersburg PA
CBHW071453220526
45472CB00003B/789